Job #: 111317

Author Name: Trudeau

Title of Book: The Non-Euclidean Revolutin

ISBN #: 9780817647827

Modern Birkhäuser Classics

Many of the original research and survey monographs in pure and applied mathematics published by Birkhäuser in recent decades have been groundbreaking and have come to be regarded as foundational to the subject. Through the MBC Series, a select number of these modern classics, entirely uncorrected, are being re-released in paperback (and as eBooks) to ensure that these treasures remain accessible to new generations of students, scholars, and researchers.

The Non-Euclidean Revolution

Richard J. Trudeau

With an Introduction by
H.S.M. Coxeter

Reprint of the 1987 Edition

Birkhäuser
Boston • Basel • Berlin

Richard J. Trudeau
Department of Mathematics
Stonehill College
North Easton, MA 02357
U.S.A.

Originally published as a monograph

ISBN-13: 978-0-8176-4782-7   e-ISBN-13: 978-0-8176-4783-4
DOI: 10.1007/978-0-8176-4783-4

Library of Congress Control Number: 2007940674

Mathematics Subject Classification (2000): 01Axx, 00A30, 03A05, 51-03, 14C40

Cover design by Alex Gerasev.

Printed on acid-free paper.

9 8 7 6 5 4 3 2 1

www.birkhauser.com

Richard J. Trudeau

The Non-Euclidean Revolution

With an Introduction by H.S.M. Coxeter

With 257 Illustrations

Birkhäuser
Boston • Basel • Berlin

Richard J. Trudeau
Department of Mathematics
Stonehill College
North Easton, MA 02357
U.S.A.

Library of Congress Cataloging in Publication Data
Trudeau, Richard J.
  The non-Euclidean revolution.
  Bibliography: p.
  Includes index.
  1. Geometry, Non-Euclidean. I. Title.
QA685.T75 1987   516.9   85-28014

CIP-Kurztitelaufnahme der Deutschen Bibliothek
Trudeau, Richard J.:
The non-Euclidean revolution / Richard J. Trudeau.
— Boston ; Basel ; Stuttgart : Birkhäuser, 1986.
  ISBN 3-7643-3311-1 (Basel ...)
  ISBN 0-8176-3311-1 (Boston)

ISBN 0-8176-3311-1
ISBN 3-7643-3311-1

Typeset by Asco Trade Typesetting Ltd., Hong Kong.
Printed and bound by R.R. Donnelley & Sons, Harrisonburg, Virginia.
Printed in the U.S.A.

9 8 7 6 5 4 3 2

To T.E. Lockary, Tom Clarke, and Bob Kruse

Preface

epistemology n. The study of the nature and origin of knowledge.

When I was in my early teens I felt adults were lying to me much of the time. They made dogmatic statements they couldn't defend. "God will punish you if you take His name in vain," they said. "A sentence must never end with a preposition.[1]" At the time what confidence I had was derived mostly from my ability to reason, so I reacted intellectually. I became obsessed with discovering what, if anything, is *true*, and in what sense.

About this time two things happened. I studied plane geometry, which I found fascinating, and I noticed that the phrase "mathematically proven" was a folk-wisdom synonym for "absolutely certain." I concluded that if absolute truth is to be found anywhere it must lie in mathematics.

In college I studied mathematics and philosophy. I learned to formulate my epistemological knot more precisely: to what extent is mathematics the truth? But I made little progress untangling it.

After college, what with graduate school, adjusting to work, and learning to like myself and trust my emotions, the knot was pushed to the back of my mind.

I became a college mathematics teacher. In the spring of 1971 my chairman asked me to develop a course for the fall term in non-Euclidean geometry. I had heard of non-Euclidean geometry but had never studied it. That summer I did study it, along with the 2100-year-old controversy that had culminated in its invention, and suddenly I was able to untangle my long-neglected epistemological knot in a most satisfactory manner. I felt that at last I understood the extent to which mathematics is true, more importantly the extent to which it is not, and by inference the extent to which any general statement in science or philosophy can claim to be true. It was a heady experience; I felt as if I had been transported to a vantage from which I could see—actually see—the limits of reason.

What I learned that summer was that a struggle with the notion of *mathematics as truth* similar to my own had unfolded in mathematical and philosophical circles from about 400 B.C. into the 19th century; that it had climaxed

vii

in the first half of the 19th century with the invention of non-Euclidean geometry; and that as a result over the second half of that century mathematicians and scientists changed the way they viewed their subjects. An entire scientific revolution had taken place that I had never heard of!

Moreover, prescinding from my special interest in the matter, I felt that this intellectual adventure I had stumbled upon made a terrific story. Thus this book; for, being a teacher, whenever I hear a good story I immediately want to retell it, in my own way, to someone else.

I presuppose that you studied plane geometry in high school. However I do not expect that you have done so recently, or that you did particularly well in the course, or that you remember much about it. And while I occasionally draw upon high school algebra to illustrate a point, if you've never studied that subject you won't be at any real disadvantage.

The book proceeds on three levels. On one it's just a geometry book with extra material on history and philosophy. For a while we will talk about Euclidean geometry—the "plane geometry" of high school—then switch to "hyperbolic geometry," another plane geometry invented around 1820. We will compare the two and reflect on what we have done.

On another level this book is about a scientific revolution, every bit as significant as the Copernican revolution in astronomy, the Darwinian revolution in biology, or the Newtonian or 20th-century revolutions in physics, but which is largely unsung because its effects have been more subtle—a revolution brought about by the invention of an alternative to traditional Euclidean geometry. Hyperbolic geometry is as logically consistent as Euclid's, has as much claim to being "true" as Euclid's, and yet extensively *contradicts* Euclid's. In Euclidean geometry the angles of a triangle add up to 180°; in hyperbolic geometry they add up to less, and the sum varies from triangle to triangle. In Euclidean geometry the Theorem of Pythagoras[2] holds; in hyperbolic geometry it does not. The effect of this paradoxical situation on 19th-century mathematicians and scientists was profound. Mathematicians embarked on an agonizing reappraisal of their subject that would last for decades; and scientists found themselves asking whether science wasn't in fact a very different thing than they had always thought.

On the third and most speculative level this book is about the possibility of significant, absolutely certain knowledge about the world. It offers striking evidence—though of course it cannot prove—that such knowledge is impossible.

I said that I assume you have studied Euclidean geometry. If in addition—and I think this is likely—you have *not* studied non-Euclidean geometry, and your epistemology of mathematics is as nebulous as mine used to be, then the story I retell in this book will provide you a rare opportunity to actually *experience* the intellectual and intuitive disorientation scientific revolutions cause. In fact the opportunity may be unique. If you are an average educated person it would probably be difficult for you, reading an account of one of the other scientific revolutions I mentioned, to feel the confusion (and excitement!) that originally surrounded the event, because *you already believe* the once-revolutionary theory to be substantially correct. You have been brought up to believe the earth moves around the sun and is held to its path by gravity (so much for Copernicus and Newton); you may have doubts about the specific mechanism Darwin proposed to *explain* evolution, but you probably con-sider it a fact that evolution has occurred, which is what the fuss was really about; and while you may not know much about 20th-century physics, the spectacle of a nuclear explosion is terrifying proof that there is something to it. With regard to geometry, however, you are almost certainly a committed Euclidean, and consider the possibility of a logical, "truthful" geometry contradicting Euclid's to be absurd. You are like a 16th-century astronomer hearing of Copernicanism for the first time.

The coming of non-Euclidean geometry was basically a mathematical event, so learning about it involves reading mathematics. Mathematics is more demanding than light fiction, so take your time. Don't try to push on when you're tired. In parts of this book you may not want to read more than two or three pages at a sitting.

On the other hand, this book is supposed to be fun. Feel free to skip parts too technical for your taste. You will be able to pick up the thread again, after the technicalities subside. Feel free, especially, to skip proofs (in Chapters 2, 4, and 6). They are the bookkeeping, included to show that matters stand as I say they do, but skippable if you'd as soon take my word for it.

I have included some exercises, in case you'd like to try your hand. If not, they too can be skipped without loss of continuity.

Richard J. Trudeau

Notes

[1] *preposition.* The record for largest number of consecutive terminal prepositions (five) seems to be held by the child's question at the end of the following anecdote, whose author I have unfortunately been unable to trace.

A child is in bed with a cold. "Mommy, can you come up and read me a story?" he calls down to his mother. As she reaches the top of the stairs, he recognizes the book in her hand as one he doesn't care for. He asks, "What did you bring the book that I don't like to be read to out of up for?"

[2] *Theorem of Pythagoras.* In every right triangle, $c^2 = a^2 + b^2$ where c denotes the length of the longest side and a and b the lengths of the other two.

Contents

Introduction

Felix Klein described non-Euclidean geometry as "one of the few parts of mathematics which is talked about in wide circles, so that any teacher may be asked about it at any moment." This old observation is now reinforced by our knowledge that astronomical space is only approximately Euclidean. Trudeau's book provides the reader with a non-technical description of the progress of thought from Plato and Euclid to Kant, Lobachevsky, and Hilbert. There is a pleasantly discursive treatment of Pontius Pilate's unanswered question "What is truth?" The text is enlivened by abundant quotations and amusing cartoons. The final chapter includes a clear account of experiments which seem to indicate that the world as seen by our two eyes is not even approximately Euclidean but hyperbolic!

H. S. M. Coxeter

CHAPTER 1

First Things

The Origin of Deductive Geometry

Once upon a time—around 600 B.C.—there was a man named Thales, who invented what we call "science."

Before Thales, thinkers did not think abstractly. Instead of looking for principles behind the curious events with which nature confronted them, they looked for personalities. Their findings—myths—were stories populated with an assortment of gods and goddesses whose interactions with one another and with human beings produced natural phenomena like spring, thunder, eclipses, etc.

Nowadays it is common to sneer at ancient myths, but they were created by men and women of genius. The myths provided a comprehensive explanation of natural phenomena and a link between humanity and nature that made the universe less frightening. In fact science has only one advantage over myth, by also predicting natural phenomena to a degree myth never could.

(By the way, this story I'm telling about pre-Euclidean geometry is itself a sort of myth. It credits a few legendary characters with subtle intellectual developments that must actually have involved numerous people over considerable time. The story has evolved, in the virtual absence of hard data, from a few legends, mathematicians' longing to know the origin of their subject, and their sense, as mathematicians, of what the milestones probably were.)

Thales (c.625–c.547) believed nature operated not by whimsy, or by the gods' romantic entanglements, but by principles intelligible to human beings. Thales introduced abstraction into the contemplation of nature.

In particular, Thales introduced abstraction into geometry. Before Thales, "geometry" had meant "surveying" (the Greek *geometrein* means "to measure land"), and geometric figures had been particular objects like corrals and fields. Instead Thales conceived of geometric figures as abstract shapes. This enabled him, when he examined in this light the hodgepodge of geometric recipes, rules-of-thumb, and empirical formulae that had been transmitted from Babylonia and Egypt,[1] to detect an order. He noticed that some geometric facts were deducible from others. And he made the extraordinary suggestion that geometry should become, as much as possible, a purely mental activity.

Greek Civilization in 550 B.C.

Thales was Greek, of course, and lived in a city that was then at the center of Greek culture: Miletos, on the western coast of Asia Minor (currently Turkey). Just a few miles from Miletos there is an island called Samos, where Pythagoras was born when Thales was in his fifties.

When Pythagoras (c.570–c.495) grew up he learned of Thales' scientific ideas. He was particularly captivated by Thales' proposal for geometry.

Pythagoras left Asia Minor, and for a while studied in Egypt. Eventually he settled in Kroton, a Greek city in southern Italy, at the ball of the foot. There he founded the "Society of Pythagoreans," a community of men and women sharing quasi-religious rituals, dietary laws, and devotion to mathematics as the key to understanding nature. Though the actual community lasted only a few decades, its doctrines continued to influence Greek thought for much longer. Centuries later various thinkers around the Mediterranean were still calling themselves "Pythagoreans" and professing the Pythagorean belief in the primacy of mathematics among the sciences. To an extent the Society has affected Western thought down to the present, for Plato (c.427–347) was strongly influenced by Pythagorean ideas. ("All philosophy is a series of footnotes to Plato"–A. N. Whitehead.)

The Pythagoreans accepted Thales' program of making geometry a deductive science. To this end their greatest contribution was a dramatic discovery that helped set the standard of proof. Thales had deduced his theorems by a combination of logic and intuitive reflection. The Pythagoreans discovered that logic and intuition can disagree!

Here's what happened. Let AB and CD be two straight line segments (see Figure 1). We will say that a straight line segment XY is a "common measure"

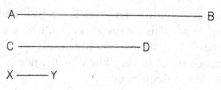

Figure 1

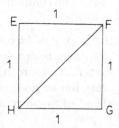

Figure 2

of AB and CD if there are whole numbers m and n so that XY laid end-over-end m times is the same length as AB and XY laid end-over-end n times is the same length as CD. For example if AB were a yard long and CD 10 inches, a segment XY of 2 inches would be a common measure with $m = 18$ and $n = 5$; for laying XY end-over-end eighteen times would produce a length of 36 inches, the same as AB, and laying XY end-over-end five times would produce a length of 10 inches, the same as CD. It was intuitively evident to the early Pythagoreans (and as I write it is intuitively plausible to me) that a common measure can be found for *any* pair of segments—though of course it may be necessary to take XY quite small in order to measure both AB and CD exactly. Since $AB/CD = (m \cdot XY)/(n \cdot XY) = m/n$, a "rational" number (that is, a ratio of whole numbers), what their intuition predicted was that the quotient of two lengths would always come out rational.

Now take a square with side equal to 1 and draw a diagonal (see Figure 2). Applying the Theorem of Pythagoras (p. ix) to the right triangle FGH we get $FH^2 = FG^2 + GH^2 = 1^2 + 1^2 = 2$, so $FH = \sqrt{2}$ and therefore the quotient FH/FG of the two lengths FH and FG is equal to $\sqrt{2}/1 = \sqrt{2}$ also. If the early Pythagoreans had been correct that the quotient of two lengths is always rational, $\sqrt{2}$ would then be rational. But one of the later Pythagoreans (probably Hippasos of Metapontion, [2] after 430 B.C.) discovered, by an argument not based (primarily) on intuition, that $\sqrt{2}$ is *not* rational.

The proof went something like this. Any rational number can be "reduced to lowest terms," that is, expressed by whole numbers having no whole number factor (other than 1) in common; for example $360/75 = 24/5$ and 24 and 5 have no common factor. Therefore if $\sqrt{2}$ were rational it would be possible to express it as $\sqrt{2} = p/q$ where p and q are whole numbers with no common factor. Squaring both sides gives $2 = p^2/q^2$, and multiplying both sides by q^2 gives $2q^2 = p^2$. This means p^2 is even, because it is twice another whole number. The Pythagoreans has previously proven that only even numbers have even squares,[3] so they knew that, since p^2 is even, p must be even also. This has two consequences:

(1) p is twice some other whole number (this is what being "even" means) which we can call "r," so $p = 2r$; and

(2) q is odd, for we said p and q have no common factor, and an even q would have a factor 2 in common with p.

We will pursue (1). Substituting $2r$ for p in the equation $2q^2 = p^2$ (above), we get $2q^2 = (2r)^2$ or $2q^2 = 4r^2$. Dividing both sides by 2 gives $q^2 = 2r^2$ so q^2, being twice a whole number, is even. As before this implies that q is even (only even numbers have even squares). But we just said in (2) that q is odd! As the hypothesis that $\sqrt{2}$ is rational has led to this contradiction, logic forces us to conclude that $\sqrt{2}$ is not rational.[4]

At this point the Pythagoreans were perplexed. They were *sure*, on intuitive grounds, that $\sqrt{2}$, being the quotient of two lengths, is a rational number. On the other hand they were equally sure, on grounds of logic and computation, that $\sqrt{2}$ is *not* a rational number!

Had the mathematical world decided to accept intuition as more reliable than logic the future of mathematics would have been quite different; but it did decide in favor of logic,[5] and mathematicians ever since have been trained to revere logic and mistrust intuition. (I think this has something to do with the generalization that mathematicians are "cold" people.)

To say mathematicians consider intuition unreliable, however, is not to say they have banished it from mathematics. On the contrary, the basic assumptions from which any branch of mathematics proceeds—the "axioms"—are accepted, without proof, primarily because of intuitive appeal. And intuition plays a big role in the discovery of theorems as well, or mathematicians would be spending most of their time trying to prove false statements. It's just that intuitive evidence is not accepted as conclusive.

The Pythagorean heritage is what modern mathematicians call "rigor," a habit of mind characteristic of mathematics. Every effort is made to insulate the subject from its down-to-earth origins. Terms are defined and principles formulated with constant vigilance against unstated assumptions. Theorems are derived by logic alone.

In the 5th century B.C., before mathematics was made rigorous, mathematicians had already constructed long chains of geometric theorems in which each theorem was deduced, informally, from those before it. Each chain started with generalizations from experience which of course were not proven.

As the scope of these chains grew there emerged the daring idea that it might be possible to link them together into a single network, anchored to a small number of generalizations from experience, which would contain a broad inventory of elementary geometric knowledge. And toward the end of the century—in fact, about the same time it was proven that $\sqrt{2}$ is not rational—a mathematician named Hippokrates of Chios[6] accomplished exactly this, in a book he called the *Elements*.

Later, while the rigorization of mathematics was underway, other comprehensive geometric networks[7] were forged. Each was called the *Elements*, and presumably each, by having simpler axioms, tighter logic, or more theorems, was an improvement on its predecessors. The series culminated in the famous *Elements* of Euclid, completed about 300 B.C.

Euclid's *Elements* is a single deductive network of 465 theorems that includes not only an enormous amount of elementary geometry, but generous helpings of algebra and number theory as well. Its organization and level of logical rigor were such that it soon became geometry's standard text. In fact it so completely superseded previous efforts that they all disappeared.

The *Elements*—from now on the title will refer to Euclid's book only—is the most successful textbook ever written. It has gone through more than a thousand editions and was used well into the last century (here and there, it is used even today). More importantly, it is the paradigm that scientists have been emulating ever since its appearance. It is the archetypal scientific treatise. To study the form and limitations of the *Elements*, therefore, is to poke through the entrails of the whole scientific enterprise.

Given the stature of this work, surprisingly little is known about its author. Scholars even hesitate to conjecture Euclid's dates, except to say that he "flourished" about 300 B.C. Just then the center of scientific and mathematical activity was shifting from Athens to Alexander the Great's new city Alexandria at the mouth of the Nile. Euclid lived in Alexandria, where he was a professor of mathematics at the Museum (the university). Beyond this all that is known about Euclid is contained in two anecdotes. In one a beginning student of geometry asks him, "What shall I get by learning these things?" Euclid responds by calling a servant and saying, "Give him a coin, since he must make gain out of what he learns." In the other the king, Ptolemy I,[8] asks him, "Is there in geometry any shorter way than the *Elements*?"—to which Euclid replies, "There is no royal road to geometry."

Material Axiomatic Systems

The *Elements* is the oldest example we possess of what is now called a "material axiomatic system." Before we examine the *Elements* itself, with all the explaining and restructuring of Euclid's text our study will involve, I think it would be wise to discuss material axiomatic systems in general (in this and the next two sections) and to illustrate Euclid's proof techniques in a nongeometric context (in the following section).

MUTT AND JEFF   **Created by Bud Fisher**

Figure 3. Courtesy of P. S. de Beaumont.

Pattern for a Material Axiomatic System[9]

(1) The basic technical terms of the discourse are introduced and their meanings explained. These basic terms are called *primitive terms*.

(2) A list of primary statements about the primitive terms is given. In order for the system to be significant to the reader, he or she must find these statements acceptable as true based on the explanations given in (1). These primary statements are called *axioms*.

(3) All other technical terms are defined by means of previously introduced terms. Technical terms which are not primitive terms are accordingly called *defined terms*.

(4) All other statements of the discourse are logically deduced from previously accepted or established statements. These derived statements are called *theorems*.

Notice there are two kinds of technical terms. The meanings of the "defined" terms (item (3)) are prescribed by reference to terms (of either type) previously introduced; at least in relation to those earlier terms, therefore, the defined terms are completely unambiguous. But unfortunately it is not possible to achieve unambiguity for *all* terms: dictionaries are, after all, circular. (See Figure 3, or try Jeff's experiment yourself with a word like "alive" or "straight" that cycles back quickly.) Thus it is necessary to accept some terms into the system (usually from everyday speech) without benefit of precise definition; these are the "primitive" terms of item (1). Of course every effort is made to indicate the sense in which each primitive term is to be taken, but no amount of explaining can guarantee that everyone will understand them in exactly the same way.

Similarly there are two kinds of statements. Just as one cannot define every term, one cannot deduce every statement. Accordingly, the statements of item (2) are accepted without deductive proof, on grounds that are outside the official structure of the system. (Within the system they are viewed simply as assumptions.) These statements provide a starting point from which all the other statements (item (4)) are logically deduced.

For many people a sticking place is that phrase, "logically deduced." Before we proceed to an example of a material axiomatic system, therefore, I think we should spend some time talking about logic—at first in general, then as we will encounter it in this book.

Logic

Every rational discussion involves the making of inferences. What *kinds* of inferences are allowed depends on who the participants are and what subject is being discussed. In this sense each type of discussion has its own special logic. For example, the sort of evidence that physicists accept as strong confirmation of a theory is rejected as totally inadequate by mathematicians trying to prove a theorem; in turn, the esoteric reasoning mathematicians sometimes employ is utterly worthless to literary critics analyzing a novel. (Indeed, there are forms of argument employed regularly in mathematics that are applicable to *nothing* else.[10])

Usually, however, the term "logic" is used in a more general sense, to refer to principles of reasoning that the various special logics are presumed to have in common. The belief is that this common logic would be acceptable and potentially useful to participants in *any* rational discussion. Of course there's no way of checking this without polling the entire planet, or at least scrutinizing its more than 3,000 languages, but since Greek concepts are so much a part of the Western heritage it seems safe to say there is a widely shared logic at least among people with Western-style educations.

Though this traditional logic does not include the special techniques of modern mathematics, it does include all the forms of argument used by mathematicians in Euclid's time. In fact, today many people, hearing the term "logic," can think of little *except* the principles of reasoning used by Euclid, because the only time they have ever heard logic discussed explicitly (rather than taken for granted) was in a high school geometry course.

Throughout this book, even when we take up non-Euclidean geometry, Euclid's logic is all we will ever need. We have good reason, therefore, to feel confident about the soundness of our logic. It is safely within traditional logic, and has been embedded in the fabric of Western thought for more than 2,000 years.

Nonetheless it is wise to take *all* logic with a grain of salt. It is vulnerable to doubt, on at least two counts.

I'll let the author of *Alice's Adventures in Wonderland* and *Through the Looking-Glass* tell you about the first.

Achilles had overtaken the Tortoise, and had seated himself comfortably on its back.

"So you've got to the end of our race-course?" said the Tortoise. "Even though it *does* consist of an infinite series of distances? I thought some wiseacre[11] or other had proved that the thing couldn't be done?"

"It *can* be done," said Achilles. "It *has* been done! *Solvitur ambulando*. You see the distances were constantly *diminishing*: and so—"

"But if they had constantly been *increasing*?" the Tortoise interrupted. "How then?"

"Then I shouldn't be *here*," Achilles modestly replied; "and *you* would have got several times round the world, by this time!"

"You flatter me–*flatten*, I mean," said the Tortoise; "for you *are* a heavy weight, and *no* mistake! Well now, would you like to hear of a race-course, that most people fancy they can get to the end of in two or three steps, while it *really* consists of an infinite number of distances, each one longer that the previous one?"

"Very much indeed!" said the Grecian warrior, as he drew from his helmet (few Grecian warrior possessed *pockets* in those days) an enormous note-book and a pencil. "Proceed! And speak *slowly*, please! *Shorthand* isn't invented yet!"

"That beautiful First Theorem of Euclid!" the Tortoise murmured dreamily. "You admire Euclid?"

"Passionately! So far, at least, as one *can* admire a treatise that won't be published for some centuries to come!"

"Well, now, let's take a little bit of the argument in that First Theorem—just *two* steps, and the conclusion drawn from them. Kindly enter them in your note-book. And, in order to refer to them conveniently, let's call them A, B, and Z:

(A) Things that are equal to the same are equal to each other.
(B) The two sides of this triangle are things that are equal to the same.
(Z) The two sides of this triangle are equal to each other.

"Readers of Euclid will grant, I suppose, that Z follows logically from A and B, so that anyone who accepts A and B as true, *must* accept Z as true?"

"Undoubtedly! The youngest child in a high school—as soon as high schools are invented, which will not be until some two thousand years later—will grant *that*."

"And if some reader had *not* yet accepted A and B as true, he might still accept the *sequence* as a *valid* one, I suppose?"

"No doubt such a reader might exist. He might say 'I accept as true the hypothetical proposition that, if A and B be true, Z must be true; but I *don't* accept A and B as true.' Such a reader would do wisely in abandoning Euclid, and taking to football."

"And might there not *also* be some reader who would say 'I accept A and B as true, but I *don't* accept the hypothetical'?"

"Certainly there might be. *He*, also, had better take to football."

"And *neither* of these readers," the Tortoise continued, "is *as yet* under any logical necessity to accept Z as true?"

"Quite so," Achilles assented.

"Well, now, I want you to consider *me* as a reader of the *second* kind, and to force me, logically, to accept Z as true."

"A tortoise playing football would be—" Achilles was beginning.

"—an anomaly, of course," the Tortoise hastily interrupted. "Don't wander from the point. Let's have Z first, and football afterwards!"

"I'm to force you to accept Z, am I?" Achilles said musingly. "And your present position is that you accept A and B, but you *don't* accept the hypothetical—"

"Let's call it C," said the Tortoise.

"—but you don't accept:

(C) If A and B are true, Z must be true."

"That is my present position," said the Tortoise.

"Then I must ask you to accept C."

"I'll do so," said the Tortoise, "as soon as you've entered it in that note-book of yours. What else have you got in it?"

"Only a few memoranda," said Achilles, nervously fluttering the leaves: "a few memoranda of—of battles in which I have distinguished myself!"

"Plenty of blank leaves, I see!" the Tortoise cheerily remarked. "We shall need them *all*!" (Achilles shuddered.) "Now write as I dictate:

(A) Things that are equal to the same are equal to each other.
(B) The two sides of this triangle are things that are equal to the same.
(C) If A and B are true, Z must be true.
(Z) The two sides of this triangle are equal to each other."

"You should call it D, not Z," said Achilles. "It comes *next* to the other three. If you accept A and B and C, you *must* accept Z."

"And why *must* I?"

"Because it follows *logically* from them. If A and B and C are true, Z *must* be true. You don't dispute *that* , I imagine?"

"If A and B and C are true, Z *must* be true," the Tortoise thoughtfully repeated. "That's *another* hypothetical, isn't it? And, if I failed to see its truth, I might accept A and B and C, and *still* not accept Z, mightn't I?"

"You might," the candid hero admitted; "though such obtuseness would certainly be phenomenal. Still, the event is *possible*. So I must ask you to grant one more hypothetical."

"Very good. I'm quite willing to grant it, as soon as you've written it down. We will call it

(D) If A and B and C are true, Z must be true."

"Have you entered that in your note-book?"

"I *have*!" Achilles joyfully exclaimed, as he ran the pencil into its sheath. "And at last we've got to the end of this ideal race-course! Now that you accept A and B and C and D, *of course* you accept Z."

"Do I?" said the Tortoise innocently. "Let's make that quite clear. I accept A and B and C and D. Suppose I *still* refuse to accept Z?"

"Then Logic would take you by the throat, and *force* you to do it!" Achilles triumphantly replied. "Logic would tell you 'You can't help yourself. Now that you've accepted A and B and C and D, you *must* accept Z!' So you've no choice, you see."

"Whatever *Logic* is good enough to tell me is worth *writing down*," said the Tortoise. "So enter it in your book, please. We will call it

(E) If A and B and C and D are true, Z must be true.

"Until I've granted *that*, of course, I needn't grant Z. So it's quite a *necessary* step, you see?"

"I see," said Achilles; and there was a touch of sadness in his tone.

Here the narrator, having pressing business at the bank, was obliged to leave the happy pair, and did not again pass the spot until some months afterwards. When he did so, Achilles was still seated on the back of the much-enduring Tortoise, and was writing in his note-book, which appeared to be nearly full. The Tortoise was saying, "Have you got that last step written down? Unless I've lost count, that makes a thousand and one. There are several millions more to come."

     —from "What the Tortoise Said to Achilles" by Lewis Carroll*

---

* *Mind*, Oxford University Press, new series, 4 (1895), pp. 278–280.

Carroll's point is that the rules of logic are not dug out of the earth like diamonds; they are grounded in human *intuition!* The imperative *feeling* we have that "If A and B are true, Z must be true" cannot be defended, or reduced further. Confronted with someone who does not share that feeling, all we can do is drop the discussion and propose football instead.

A present-day mathematical logician (Rosser, in *Logic for Mathematicians*, McGraw-Hill, 1953, p. 11) makes the same point as follows.

[The mathematician] should not forget that his intuition is the final authority, so that, in case of irreconcilable conflict between his intuition and some system of ... logic, he should abandon the ... logic. He can try other systems of ... logic, and perhaps find one more to his liking, but it would be difficult to change his intuition.

Much as the mathematician would like to seal his system off from intuition, which he considers unreliable, core intuitions penetrate every barrier. Logic itself rests on intuition, and may be contaminated with intuition's unreliability.

We learn logic, at least informally, along with our Western languages. In this sense logic is like a pair of tinted eyeglasses with which we are fitted early in life, of which we are barely aware, and through which we become, by the standards of our culture, intellectually mature. They color everything, so naturally we tend to see confirmation wherever we look. But do they also distort? Could common logic be somehow in error? (Could millions of people be wrong?) We naturally tend not to think so. But conceivably, yes. For all we know there may be something "wrong" with our languages, or even our brains. It might be that the only people who reason correctly are a few outcasts playing football!

The second count on which logic is vulnerable to doubt is the presence—and persistence—of logical paradoxes. These are arguments that, seemingly, start from sound premises, violate none of the rules of logic, but reach conclusions that are erroneous or contradictory.

To be sure, some of the paradoxes that have been announced over the years have been resolved to more or less everyone's satisfaction. Such is the case, for example, with Zeno's paradox of Achilles and the Tortoise (see Note 11 at the end of the chapter), which is usually explained as being the result of an assumption which mathematicians now know to be false: that the sum of infinitely many magnitudes must always be infinitely large.

But other paradoxes have been more troublesome. A whole rash of them announced around the turn of this century, for example (see Mendelson, *Introduction to Mathematical Logic* (Van Nostrand, 1979), pp. 2–4), have not been so much "resolved" by logicians as "avoided"—by ordaining new logical rules that prohibit the reasoning on which the paradoxes depend. This course of action is widely regarded as merely a stopgap while research for more satisfactory settlements continues.

Carroll's point was: we can't be sure that logic is *right*. Unresolved paradoxes go further: they seem to show that logic is *wrong*. Therefore, not surprisingly, each new paradox is afforded the closest attention. Here is a more recent example.*

"A new and powerful paradox has come to light." This is the opening sentence of a mind-twisting article by Michael Scriven that appeared in the July 1951 issue of the British philosophical journal *Mind* . . . . That the paradox is indeed powerful has been amply confirmed by the fact that more than twenty articles about it have appeared in learned journals. The authors, many of whom are distinguished philosophers, disagree sharply in their attempts to resolve the paradox. Since no consensus has been reached, the paradox is still very much a controversial topic.

. . . It often takes the form of a puzzle about a man condemned to be hanged.

The man was sentenced on Saturday. "The hanging will take place at noon," said the judge to the prisoner, "on one of the seven days of next week. But you will not know which day it is until you are so informed on the morning of the day of the hanging."

The judge was known to be a man who always kept his word. The prisoner, accompanied by his lawyer, went back to his cell. As soon as the two men were alone the lawyer broke into a grin. "Don't you see?" he exclaimed. "The judge's sentence cannot possibly be carried out."

"I don't see," said the prisoner.

"Let me explain. They obviously can't hang you next Saturday. Saturday is the last day of the week. On Friday afternoon you would still be alive and you would know with absolute certainty that the hanging would be on Saturday. You would know this *before* you were told so on Saturday morning. That would violate the judge's decree."

"True," said the prisoner.

"Saturday, then is positively ruled out," continued the lawyer. "This leaves Friday as the last day they can hang you. But they can't hang you on Friday because by Thursday afternoon only two days would remain: Friday and Saturday. Since Saturday is not a possible day, the hanging would have to be on Friday. Your knowledge of that fact would violate the judge's decree again. So Friday is out. This leaves Thursday as the last possible day. But Thursday is out because if you're alive Wednesday afternoon, you'll know that Thursday is to be the day."

"I get it," said the prisoner, who was beginning to feel much better. "In exactly the same way I can rule out Wednesday, Tuesday, and Monday. That leaves only tomorrow. But they can't hang me tomorrow because I know it today!"

. . .

. . . The prisoner is convinced, by what appears to be unimpeachable logic, that he cannot be hanged without contradicting the conditions specified in his sentence. Then on Thursday morning, to his great surprise, the hangman arrives. Clearly he did not expect him. What is more surprising, the judge's decree is now seen to be perfectly correct. The sentence can be carried out exactly as stated. "I think this flavour of logic refuted by the world makes the paradox rather fascinating," writes Scriven. "The logician goes pathetically through the motions that have always worked the spell before, but somehow the monster, Reality, has missed the point and advances still."

---

* Reprinted from "Mathematical Games," Martin Gardner, from *Scientific American*, March 1963, with permission of W. H. Freeman and Co.

Whether or not this paradox is ever resolved, the plight of the prisoner is an interesting metaphor for what *could* be our situation as logical thinkers. It might be, as the Argentinian poet Jorge Luis Borges expresses it in an essay inspired by the Carroll story, that

We ... have dreamt the world. We have dreamt it as firm, mysterious, visible, ubiquitous in space and durable in time; but in its architecture we have allowed tenuous and eternal crevices of unreason [the logical paradoxes] which tell us it is false.

—from "Avatars of the Tortoise" *

Proofs[12]

Interesting as it might be to speculate further about the reliability of logic, that is not the main business of this book. Accordingly let us accept the correctness of traditional logic (as we are strongly inclined to do), return our attention to item (4) of the Pattern for a Material Axiomatic System (page 6), and discuss the *manner* in which logical deduction of theorems is actually carried out by the mathematicians whose work we will be studying.

Theorems are "conditional" statements, that is statements of the form "If ..., then ...." The part between "if" and "then" is the "hypothesis," and the part after "then" is the "conclusion." In some cases a theorem is written in a form that does not appear to be conditional, but nonetheless it can be translated into that form. For example

In isosceles triangles the angles at the base are equal to one another

(Euclid's Theorem I.5) can be rewritten

If a triangle has two equal sides, then the angles subtended by those sides are equal.

From the second version we see that the hypothesis is "a triangle has two equal sides" and the conclusion is "the angles subtended by [i.e., opposite] those sides are equal."

A theorem is not *merely* a conditional statement; it is a conditional statement having a *proof*. This brings up the main point.

A "proof" is a list of statements, with a justification for each, ending with the desired conclusion. *Only seven types of justification are allowed*:

(1) "by hypothesis"   (the hypothesis of the theorem)
(2) "by RAA hypothesis"   (in case of proof by contradiction—see below)
(3) "by Axiom ..."   (an axiom of the system)

---

* Jorges Luis Borges: *Labyrinths*. Copyright © 1962, 1964 by New Directions Publishing Corporation. Reprinted by permission of New Directions.

(4) "by Theorem ... "   (a previously-proven theorem)
(5) "by Definition ... "   (the meaning of a *defined* term)
(6) "by step ... "   (an earlier step in the proof)
(7) "by rule ... of logic"   (a rule of logic)

In proving a theorem one accepts as true the hypothesis, the axioms, and all previously-proven theorems, and tries to deduce the conclusion.

Note that one may appeal *only* to the meaning of a defined term and *never* to the meaning of a primitive term. Mathematicians insist on this because there are subtle differences in people's understandings of the primitive terms. My intuition may invest a primitive term with a certain property and your intuition may differ; if I use that property in a proof you will find my proof incomprehensible. This is bad, for a proof is supposed to be a communication between people. Hence we insist that the only properties of primitive terms that may be used in a proof are those properties we agreed on at the outset— namely, those mentioned in the axioms. If we wish to assert a property of a primitive term not specified in an axiom, we must either prove a theorem that says the term has the property or, failing that but agreed we want the primitive term to have the property, add a new axiom that provides the primitive term with the desired property.

We will mention very few rules of logic by name. A complete list would swell this part of the book out of all proportion. If you happen to have studied logic formally you will have no trouble assigning the usual names, if you care to, to the forms of inference we will employ. If you have not studied logic formally— and I assume my readers have *not*—rest assured that you will not be at the slightest disadvantage. Any tricky reasoning will be carefully explained (for everyone's sake), and the rest will be as straightforward as deducing Z from A and B in Carroll's syllogism

(A) Things that are equal to the same are equal to each other.
(B) The two sides of this triangle are things that are equal to the same.
(Z) Therefore, the two sides of this triangle are equal to each other.

For a simple deduction like this—and most of our deductions will be equally simple—our standard will be that "by steps A and B" is an adequate reason for step Z. The validity of our spontaneous judgment that "If A and B are true, Z must be true" is not diminished by our failure to mention the official rule under which a logician would classify it.

Two rules of logic we *should* mention explicitly provide the basis for "proof by contradiction," a popular mathematical technique some people have trouble understanding.

Law of the Excluded Middle: Either a statement is true or its negation is true (there is no "middle" possibility).
Law of Contradiction: A statement and its negation cannot both be true.

Proof by contradiction is also called "indirect proof" or "proof by *reductio ad absurdum* [reduction to absurdity]." Concerning this technique the accomplished British mathematician G.H. Hardy[13] (1877–1947) wrote

> [R]eductio ad absurdum, which Euclid loved so much, is one of a mathematician's finest weapons. It is a far finer gambit than any chess gambit: a chess player may offer the sacrifice of a pawn or even a piece, but a mathematician offers *the game*.
>
> —*A Mathematician's Apology*, p.94

To prove a theorem "If H, then C" (H for hypothesis, C for conclusion) by this technique, we begin by assuming that $\sim$C (the negation of C) is true. This opposite assumption is called the "RAA [for *reductio ad absurdum*] hypothesis." It is a *temporary* assumption from which we derive a *contradiction*. Once we have shown that $\sim$C leads to a contradiction it follows from the Laws of Contradiction and Excluded Middle that C must be true.

A "contradiction" is a statement that denies something accepted as true: an axiom, a previous theorem, the hypothesis H, or even the RAA hypothesis $\sim$C itself.

Proof by contradiction amounts to saying, "Consider the alternative." I have H, I want to show C. I know by the two laws of logic that one and only one of the two statements, C or $\sim$C, is true, so the only alternative to C being true is that $\sim$C is true. So I investigate the possibility that $\sim$C is true. In deducing a contradiction I discover that if $\sim$C were true then something silly would happen. So I conclude that $\sim$C is unacceptable and that consequently C is the statement that is true.

The Pythagorean proof that $\sqrt{2}$ is not a rational number (pp. 3–4) is a proof by contradiction. Here the hypothesis H is that $\sqrt{2}$ is a number whose square is equal to 2. (H is usually not mentioned because we take it for granted—it is implicit in our understanding of the symbol "$\sqrt{2}$.") The desired conclusion C is that the number $\sqrt{2}$ is not a *rational* number (a ratio of whole numbers). The other possibility, $\sim$C, is of course that $\sqrt{2}$ *is* a rational number. The proof begins by entertaining the possibility that $\sim$C is true, but ends by concluding that as this leads to a contradiction C must be true instead. The argument proceeds, in outline, as follows. (See our earlier presentation for details.)

1. Pretend that $\sqrt{2}$ is rational. (This is $\sim$C, the RAA hypothesis.)

2. Then there are whole numbers p and q with no common factor (other than 1) such that $\sqrt{2} = p/q$. (This is by step 1, along with the definition of "rational" number and a previous theorem that says every rational number can be reduced to lowest terms.)

3. Then $2q^2 = p^2$. (Deduced from step 2 using the hypothesis H and previous theorems—the rules of arithmetic.)

4. Then p^2 is even. (Deduced from step 3 by the definition of the term "even.")

5. Then p is even. (By step 4 and a previous theorem that says only even numbers have even squares.)

6. Then q is odd.	(By step 2 (p and q have no common factor), step 5, and the definitions of "even" and "odd.")
7. q^2 is even.	(Deduced from steps 5 and 3 by the rules of arithmetic and the definition of "even.")
8. Then q is even.	(From step 7 and the theorem used in step 5.)
9. Contradiction.	(In view of the definitions of "even" and "odd," step 8 conflicts with step 6.)
10. Therefore $\sqrt{2}$ is not rational.	(As $\sim$C has led to an absurdity and C is the only alternative, we conclude that C is the statement that is true.)

I think there are two reasons why the method of proof by contradiction tends to confuse people. First, it seems ludicrous to begin a proof by *pretending*. How can a fantasy lead to anything solid? And second, the jump from the step announcing the contradiction to the concluding step (above, the jump from step 9 to step 10) can be psychologically enormous, because the intricacies under discussion just prior to the contradiction announcement often have no obvious relation to the conclusion (above, then evenness of q^2 or q seem to have nothing to do with the nonrationality of $\sqrt{2}$).

The remedy is to realize that in pretending $\sim$C is true (above, in pretending $\sqrt{2}$ is rational) we are not agreeing that $\sim$C *is* true, but only to *entertain the possibility* that it *might* be true, in order to see what would happen if it were; and that the ensuing contradiction signifies that, in fact, it is not. (Above, the absurdity of a number both even and odd is what *would* happen if $\sqrt{2}$ *were* rational, so we conclude that it is *not* rational.)

Today there are minority schools of mathematicians, the Intuitionists and Constructivists, who in some contexts reject the Law of Excluded Middle and therefore the unrestricted use of proof by contradiction. Their criticism has not, however, had much impact on mainstream mathematics. In any case, since proof by contradiction was not seriously questioned until *after* the invention of non-Euclidean geometry, and the mathematicians *we* will be studying used proof by contradiction without the slightest hesitation (and *often*), we will freely use this technique whenever it seems advantageous to do so.

A Simple Example of a Material Axiomatic System[14]

This system is called "The Turtle Club." Compare it with the Pattern on p. 6.

The primitive terms are *person* and *collection*. It is intended that these terms be understood in their everyday senses.

Definitions. The *Turtle Club* is a collection of one or more persons. A person in the Turtle Club is called a *Turtle*. The *Committees* are certain collections of one or more Turtles. A Turtle in a Committee is called a *member* of that Committee. Two Committees are *equal* if every member of the first is also a member of the second and every member of the second is also a member of the first. Two Committees having no members in common are called *disjoint* Committees.

Axioms.

(1) Every Turtle is a member of at least one Committee.
(2) For every pair of Turtles there is one and only one Committee of which both are members.
(3) For every Committee there is one and only one disjoint Committee.

The Pattern on page 6 says that Axioms must be "statements about the primitive terms"—which these are, ultimately, though defined terms are used to state them economically—that are "acceptable as true" to the reader. *I* find Axioms (1)–(3) acceptable because they accurately describe a list of names and committee assignments I have before me as I write. If you will agree to accept my testimony, we can get on with the business of proving theorems. (This is unorthodox, I know. When we get to Euclidean geometry you will have the testimony of your own common sense.)

Theorem 1. *Every Turtle is a member of at least two Committees.*

Restatement in conditional form: "If a person is a Turtle, then that person is a member of at least two Committees." Often the conditional form of a theorem is awkward.

Proof (A commentary follows).

Statements	Reasons
1. Let "*t*" be a Turtle (see Figure 4).	hypothesis, naming
2. *t* is a member of some Committee "*C*".	Ax 1, naming
3. Let "*D*" be the Committee which is disjoint with *C*.	Ax 3, naming

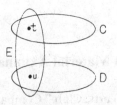

Figure 4

4. Let "*u*" be a member of *D*.   Def "Committee,"
                                                    naming
5. *u* is not a member of *C*.   Def "disjoint"
6. There is a Committee "*E*" of which *t* and *u* are   Ax 2, naming
   members.
7. *C* and *E* are not equal.   Def "equal," 5, 6
8. *t* is a member of both *C* and *E*.   2, 6
9. *t* is a member of at least two Committees.   7, 8
10. Therefore, every Turtle is a member of at least   generalization
    two Committees.

Figure 4 is *not* part of the proof—mathematical drawings never are. It does, however, make the reasoning easier to follow. I have represented Turtles by dots and Committees by ovals.

In step 1 I select and name a typical Turtle. I know that such a person exists because that is given by the hypothesis (also by the Definitions of "Turtle Club" and "Turtle"). It is a rule of logic (to which I refer by writing "naming") that anything known to exist can be named for future reference.

Though *t* is a *typical* Turtle notice that I have not specified *which* Turtle (which person on my list) *t* is. This is because I really want to talk about *all* Turtles *at once*. If I reason about a typical Turtle and use *only* characteristics that person shares with *all other* Turtles, my deductions will apply to *all* Turtles. There is a rule of logic (called "generalization") that says this is valid. I invoked it in step 10, having shown that my representative Turtle *t* is a member of at least two Committees. In everyday usage not every individual is "typical" and "generalizations" are often wrong; however in mathematics "typical" means "having properties shared by every individual without exception," making mathematical "generalizations" completely reliable.

The crucial thing about naming is that I must *have* something in order to name it. Naming presumes existence. That's why in step 2 I first invoke Axiom 1, which says that indeed there is a Committee to which *t* belongs. Step 1 is similar in this respect, as we have seen; so, too, are steps 3, 4, and 6.

In step 4 I refer to the Definition of "Committee" to support my implicit assertion that Committee *D* has at least one member. "How silly!" you might say. "Of course *D* has a member. A Committee is basically a collection of persons. Who ever heard of a collection of persons containing no person?" *You* may not have, but can you be sure others understand "collection" in the same way? "Collection" is a primitive term and we said a few pages ago that the meaning of a primitive term cannot be used in a proof. So I refer to the Definition of "Committee" instead. (By the way, mathematicians talk about *empty* "collections" all the time. It is part of their "everyday sense" of the word.)

In future proofs I will omit writing "naming," because that logical rule is used a lot and I don't want our reasons to become cluttered. I will retain, however, the practice of placing a name in quotation marks the first time it is used. That will be your signal to ask yourself, "How do we know this thing we are naming exists?" and to look for an answer among the reasons.

This brings up a more general point. I do not intend our reasons to include every single thing on which the statements depend. Step 5, for example, depends on steps 3 and 4 as well as on the Definition, so a *complete* reason would be "3, 4, Def 'disjoint'." The reasons for some other steps are similarly incomplete. My general policy will be to indicate, for each step, only the principal items on which that inference depends, omitting details I feel you could easily supply (should you care to) but which, if included, would tend to obscure the main line of argument. But I would rather risk redundancy than unintelligibility, so whenever a step is a bit tricky, like step 7, I will provide extra information.

Theorem 2. *Every Committee has at least two members.*

Restatement in conditional form: "If a collection of Turtles constitutes one of the Committees, then the collection contains at least two Turtles." In the future I won't restate a Theorem unless its original form makes it particularly hard to pick out the hypothesis and conclusion.

Proof

1. Let "C" be one of the Committees (see Figure 5).	hypothesis
2. C has at least one member "t".	Def "Committee"
3. Pretend that t is the only member of C.	RAA hypothesis
4. t is a Turtle.	Def "Committee"
5. t is a member of a second Committee "D".	Th 1
6. Let "E" be the Committee disjoint with D.	Ax 3
7. t is not a member of E.	Def "disjoint," 5, 6
8. C and E are disjoint.	Def "disjoint," 3, 7
9. E is disjoint with both C and D.	6, 8
10. But E is disjoint with only *one* Committee.	Ax 3
11. Contradiction.	9 and 10
12. Therefore C has at least two members.	3–11, logic
13. Therefore, every Committee has at least two members.	generalization

The proof is by contradiction. I want to show that C has at least two members but no direct argument occurs to me. So in step 3 I take the easy

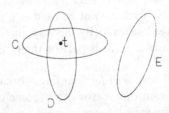

Figure 5

course and pretend that t is the only member of C. I say "easy" course because in a proof by contradiction I don't need to know where I'm going. *Any* contradiction will do, and eventually one turns up. Proof by contradiction is the scatter-gun approach to deduction.

The reason I give for step 12 is designed to remind you of how proof by contradiction works. Steps 3–11 show that if t *were* the only member of C (step 3), a contradiction (step 11) would ensue, and so, by the Laws of Contradiction and Excluded Middle ("logic"), t must *not* be the only member of C.

Proofs by contradiction are often easier to devise than direct proofs. The RAA hypothesis supplies an extra "given," and there is no particular destination, just a contradiction. Direct proofs, on the other hand, are generally more *illuminating*. If an important theorem has been proven only by contradiction, therefore, mathematicians frequently continue the search for a direct proof. Because they do not always succeed, Intuitionist and Constructivist criticism of proof by contradiction cannot be ignored.

Exercises

1. Call a whole number *trine* if it is a multiple of 3, so the "trine" numbers are 3, 6, 9, 12, 15, and so on. Prove that only trine numbers have trine squares. Use this to prove that $\sqrt{3}$ is not rational.

2. Prove that every Turtle is a member of at least *three* Committees.

3. What is the smallest possible number of Turtles? Of Committees?

Notes

[1] *Babylonia and Egypt.* In Thales' time both civilizations had long mathematical traditions, and had amassed considerable geometric and arithmetic knowledge. While the extent to which this mathematics was already abstract or deductive when the Greeks acquired it is a matter of debate, it was certainly not organized into deductive *systems* such as the Greeks would shortly construct.

[2] *Hippasos of Metapontion* is said to have been set apart from his fellow Pythagoreans because of his liberal political views. Beyond this and his discovery that $\sqrt{2}$ is not rational, all that is rumored of him is that he studied the properties of the regular dodecahedron (Figure 6).

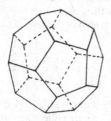

Figure 6

[3] *only even numbers have even squares.* Another way of saying this is to say that the square of an odd number is odd. Here's an algebraic proof of the latter. An even number is a whole number that is twice some other whole number; algebraically, an even number is a number of the form $2k$, where k is a whole number. An odd number can be defined as a number that is 1 less than an even number, so algebraically an odd number is a number of the form $2k - 1$, where k is a whole number. To prove that the square of an odd number must be odd, then, we have to verify that squaring a number of the form $2k - 1$ yields a result of the same shape. The square of $2k - 1$ is $4k^2 - 4k + 1 = 4k^2 - 4k + 2 - 1 = 2(2k^2 - 2k + 1) - 1$. That this last is odd can be seen directly by noting that it is 1 less than the even number $2(2k^2 - 2k + 1)$ (which is even because it is twice the whole number $2k^2 - 2k + 1$); alternatively, we can let a new symbol, say t, stand for the whole number $2k^2 - 2k + 1$, transforming $2(2k^2 - 2k + 1) - 1$ into $2t - 1$ which, having the form of an odd number, is therefore odd.

[4] $\sqrt{2}$ *is not rational.* The irrationality of $\sqrt{2}$ means that $\sqrt{2}$ is *inexpressible* in terms of whole numbers and a finite number of additions, subtractions, multiplications, and divisions. It is amazing that with such simple tools (the reducibility of rational numbers to lowest terms, the distinction between even and odd, the fact that only even numbers have even squares) it is possible to derive such a mathematically profound result.

[5] *it did decide in favor of logic.* When the mathematical world decided in favor of logic is, however, a matter of considerable debate. Decades may have passed before a consensus was reached. Plato was criticizing mathematicians for their *lack* of logically rigorous standards well into the 4th century B.C.

[6] *Hippokrates of Chios.* Chios is an island in the Aegean, near Samos, where Hippokrates was born. His birthplace is always mentioned with his name to avoid confusing him with his better-known contemporary, the "father of medicine" Hippokrates of Kos (Kos is another island in the Aegean), whose ideals are preserved in the Hippokratic Oath once sworn by the graduates of medical schools. *Our* Hippokrates probably studied mathematics on Chios then spent much of his later life in Athens, which was then becoming the center of mathematical activity. He seems to have been at least sympathetic toward the Pythagoreans, and may have been one himself. He "flourished," as historians say, about 430 B.C., meaning their best guess is he was then in his prime.

[7] *other comprehensive geometric networks.* One *Elements* was by a mathematician named Leon, of whom we know nothing beyond his name and that he flourished about 380 B.C. Another, a bit later, was by Theudios of Magnesia, who was a member of Plato's Academy.

[8] *Ptolemy I.* Ptolemy Soter, former general under Alexander the Great and first of a line of Greek rulers of Egypt that ended with Cleopatra VII (the famous Cleopatra) in 30 B.C. Not to be confused with Claudius Ptolemy, the astronomer, who worked in Alexandria around A.D. 150.

[9] *Pattern for a Material Axiomatic System.* From Eves, *A Survey of Geometry* (Allyn and Bacon, 1972), p. 11.

[10] *applicable to nothing else.* An example is the Principle of Complete Induction: Any statement that is true for the number 1 and, whenever true for a positive whole number, is also true for the next whole number, is true for every positive whole number. This is used to prove, for example, that the statement $1 + 3 + \cdots + (2n - 1) = n^2$ is true for every positive whole number n, which is the same as saying that the pattern

$$1 = 1^2$$
$$1 + 3 = 2^2$$
$$1 + 3 + 5 = 3^2$$
$$1 + 3 + 5 + 7 = 4^2$$

continues forever. The Principle is applicable only in mathematics because no other subject makes statements about infinite collections of objects.

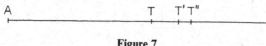

Figure 7

[11] *wiseacre.* The Tortoise refers to Zeno of Elea (flourished c. 450 B.C.), author of a number of famous paradoxes that bear his name, the implications of which are still debated. One of these, the Paradox of Achilles and the Tortoise, runs as follows (from Ettore Carruccio, *Mathematics and Logic in History and in Contemporary Thought* (Faber and Faber, 1964), pp. 33–34. Reprinted with permission.)

Swift Achilles and the slow Tortoise agree to run a race. The paradox consists in proving that if Achilles gives the Tortoise a start he will not be able to catch up with him. Let Achilles be at A and the Tortoise at T. The race is run along the straight line [see Figure 7]. ...

When he is given the signal at the beginning of the race, Achilles dashes from A to T faster than it takes to say it, but when he arrives there the Tortoise is no longer at T, but a little further ahead, at T'. Achilles reaches T', but the Tortoise is no longer at T'; he is a little further ahead at T'', and so on, to the infinite. To catch up with the Tortoise Achilles must cover the infinite segments AT, TT', $T'T''$, ... ; to run infinite segments would take an infinite time, and therefore he will never be able to catch up with the Tortoise.

Mathematicians have [an] answer: it is true that the segments AT, TT', $T'T''$, ⋯ are infinite in number, but their sum is a finite segment that can be run in a finite time and calculated as the sum of an infinite geometrical progression of common ratio less than 1. If AT is equal to 100 times the unit of measure, and if the speed of Achilles is ten times the speed of the Tortoise, the segment Achilles must cover to catch up with the Tortoise can be measured by:

$$S = 100 + 10 + 1 + \frac{1}{10} + \cdots = \frac{100}{1 - (1/10)} = \frac{1000}{9} = 111\frac{1}{9}.$$

Carroll's story opens just after Achilles has caught up with the Tortoise.

[12] *Proofs.* Some of this article is adapted from Greenberg, *Euclidean and Non-Euclidean Geometries* (Freeman, 1980), pp. 32–34.

[13] *G. H. Hardy.* In the 1920s Hardy was, by his own estimate, the fifth best pure mathematician in the world. In later life, his power to create new mathematics waning, he turned to writing first-rate textbooks and the wonderful *A Mathematician's Apology*, a defense of his life's work, largely on aesthetic grounds (Cambridge University Press, 1940, 1967). Of the *Apology* novelist and playwright Graham Greene wrote in a review,

I know no writing—except perhaps Henry James's introductory essays—that conveys so clearly and with such an absence of fuss the excitement of the creative artist.

Since 1967, reprintings of *A Mathematician's Apology* have included a touching foreword by Hardy's longtime friend, author and physicist C. P. Snow.

[14] *Example of a Material Axiomatic System.* Adapted from Eves, *op. cit.*, p. 15.

CHAPTER 2

Euclidean Geometry

As I remarked in the Preface, I assume you studied plane geometry in high school. I don't expect that you remember the details, but I do hope you retain some feeling for how the game is played.

We are about to examine Book I of the *Elements*. (In all, the *Elements* has XIII "books," though today they would be called "chapters.") As all high school texts are based at least indirectly on the *Elements*, most of the theorems will be familiar. However, our purpose is not so much to study the content of Euclid's work as to analyze its structure.

Before we begin I should warn you that whenever I say "Euclid says..." or "Euclid assumes..." or the like, I am only speaking metaphorically. No one knows, exactly, what Euclid actually wrote, let alone thought, because the text we have today is the product of many hands.

Euclid's *Elements* was soon established as the standard introduction to geometry, and copies were much in demand. As every copy was handmade, even direct copies of the original manuscript must have differed somewhat from each other. Changes in the text could only accumulate as these copies, and copies of these copies, were distributed around the Mediterranean, and were copied and recopied in their turn, and so on over the centuries. Sometimes changes were made deliberately, as when Theon of Alexandria[1] (4th century A.D.), displeased with the version that had come down to him after almost 700 years, clarified the language, interpolated steps in proofs, and added alternate proofs and minor theorems that were entirely his own.

The first printed version of the *Elements* was descended from Theon's revision, as follows. About 400 years after Theon, a copy (or a copy of a copy, or a copy of a copy of a copy, etc.) of Theon's revision was translated into Arabic. Then, about 1120, a copy (or a copy of a copy, etc.) of the Arabic translation was translated into Latin by the English philosopher Adelard of Bath. Then, about 1270, Adelard's translation (or a copy, etc.) was revised, in light of other Arabic sources (themselves derived from possibly different Greek versions of Theon's revision), by the Italian scientist Campanus of Novara. Finally, Campanus's revision (or a copy, etc.) was printed in Venice in 1482. Though the title page said the work was Euclid's, untold alterations had been made on the roughly 1800-year voyage from Euclid's hand into print.

Since 1482 a number of *Greek* versions of Theon's revision have come to light and, miraculously, one Greek *Elements* not based on Theon's revision and presumably based on an older version. From these sources a new Greek text was compiled in the 1880s by the Danish philologist Johan L. Heiberg, which is probably the closest scholars will ever come to reconstructing the original. Our quotations of the *Elements* are from this text's 1908 translation into English by Sir Thomas L. Heath.[2]

How Big Is a Point?*

The *Elements* has no preface or introduction, no statement of objectives; it offers no motivation or commentary. It opens abruptly with a list of 23 "Definitions" at the beginning of Book I. Of these, we will need 19, of which the first is

Definition 1. A *point* is that which has no part.

Euclid is saying that a point cannot be divided into parts, even in thought; it has neither length, nor width, nor thickness.

When a modern physicist studies a pendulum, he or she makes simplifying assumptions that all its mass is concentrated in the bob, that there is no friction at the pivot, and that it swings in a perfect vacuum. Of course no such pendulum exists. A physicist's pendulum is an *ideal* pendulum, the conceptual limit approached by real pendulums with progressively less massive wires and better-lubricated pivots, swinging in rarer and rarer atmospheres.

We can't be sure what Euclid was thinking as he composed the *Elements*, because the text is devoid of explanatory passages; but it appears he saw geometry as something akin to modern physics. A "point" is an idealization of a dot much as a physicist's pendulum is an idealization of a real pendulum. A point is the limit approached by dots that become smaller and smaller.

It's all right to imagine a point as a real dot whose dimensions are negligibly small—so that in *effect* it "has no part"—as long as we remember that in fact it is a ideal object whose length, width, and thickness are absolutely zero.

At this juncture your common sense might object.

"Don't you have that backward?" I imagine someone saying. "I can understand how a point's diameter—I'm thinking of a point as a tiny ball—might be so small that we can ignore it in practice, the way a chemist would ignore the diameter of an atom. But if a point were truly to have no size at all, how

---

* The following article and Figure 8 by the author and sketch by K. Branigan first appeared in *Two-Year College Mathematics Journal*, September 1983. Reprinted here with permission of The Mathematical Association of America.

could even an infinite number of them make up a line segment one meter long? No matter how many zeroes you add up, I can't see how the total could be anything but zero."

I can't see how the total could be anything but zero, either, but it doesn't bother me as much as it probably does you.

"I should think it would bother you more. You're the mathematician, and it's a mathematical argument."

Not really, though I admit it *seems* to be a mathematical argument. It's really more intuitive than logical.

"How can you call it intuitive? Look. We have a line segment one meter long—"

Yes.

"—and this segment is made up of points, laid end to end—"

Be careful. If points have no size, how could they have "ends"? And what would it mean, then, for points to be "laid end to end"? Do you see what I'm getting at?

"Sort of ... but wouldn't that mean the notion of points without size has gotten us into hot water even sooner than I had thought?"

Intuitive hot water. We can't conjure up a detailed image of how, exactly, points make up a line. But not logical hot water, at least not obviously. We have run up against a failure of our power to imagine, which makes the discussion a little strange; but as yet there is no clear-cut contradiction.

"... we have a line segment one meter long. It is *somehow* made up of points—we won't worry how—"

Good. That's exactly the attitude a mathematician would—

"But! You're saying each of those points has length *zero*—"

That's Euclid's Definition 1, yes.

"But if each of those points contributes *zero* to the one-meter length, then the entire segment must have length zero as well! *There*'s your contradiction."

From the fact that each point has length zero, why does it follow that the entire segment has length zero?

... [Frustrated silence.]

It must seem like I'm quibbling, but honestly I'm not. This issue is very subtle, very difficult to disentangle. I'd like to persuade you that the position Euclid takes in Definition 1 is the *only* one logically open to him. I need your help, though. Tell me: why, in your mind, from the fact that each point has length zero, does it follow that the entire segment has length zero? Say it as carefully as you can.

"... because the segment is made up of points *exclusively*. All its qualities must derive from those of the points. Its length, in particular, must come from the lengths of the points.

"I still want to say the length of the segment is simply the *sum* of the lengths of the points, because I feel you *were* quibbling when you objected to my description of the points as 'laid end to end.' They are, in *some* fashion, 'lined up' to form the segment. Its length must be the sum of theirs and is, therefore, zero.

"But even if your objection has substance, the fact remains that the length of the segment is *somehow* produced by the lengths of its points, and I can see no way of mathematically combining a lot of zeroes to get anything but another zero."

OK, good.

Let me begin by saying that my own common sense objects to Euclid's Definition 1 just as strenuously as yours does, and has done so continually since I first studied geometry in high school. But I have learned to ignore it.

That may seem odd—how could one ignore one's own common sense? Einstein once said[3] common sense is actually nothing more than a deposit of prejudices laid down in the mind prior to the age of eighteen. My own view is not so harsh, but I do think of common sense as consisting of *less* than a person's entire intellectual apparatus, and so limited. Without attempting to define "common sense" precisely, let me say that when I use the term in a mathematical context, I understand it to include one's powers of intuition and imagination, but not those of logic or computation. Of course the elaboration of common sense can *involve* logic or computation, as when a judgment made by the intuition or imagination is used as a premise in a deduction or a datum for a computation. But I see logic and computation as essentially distinct from common sense, because the raw material on which logic and computation operate can have other sources as well—a list of axioms, for example—in which case the conclusions are independent of intuition and imagination, and can even be opposed to them.

What I have in mind, in other words, is a distinction between two types of reasoning: "common sense," which accepts material—be the amount ever so tiny—from the intuition or imagination; and "mathematical reasoning," which accepts none whatsoever. And while the history of mathematics teaches that what one generation considers the soul of mathematical reasoning may be jettisoned by a later generation as hopelessly intuitive, making it sometimes appear that mathematical reasoning is more a goal than an accomplishment, it is the struggle for precisely this goal that has been, since the time of the Pythagoreans, the hallmark of mathematics.

Common sense can be ignored because it is not the only mode of thinking we have. And in mathematics it *must* be ignored when it conflicts with logic or computation.

"May I butt in?"

Of course. I got carried away.

"I think I see where you're headed. You're going to say that my position is based on common sense, that it conflicts with logic, and that Euclid's position is the only alternative. Am I right?"

Yes.

"Then I have two problems. First, I don't see how my position is based on intuition. Second, I don't see how it leads to a logical contradicton—in fact, it seems to show that *Euclid*'s position leads to a logical contradiction."

You base your position on intuition when you insist that the length of a segment is some mathematical combination—you lean toward the sum—of the lengths of its constituent points. But there are some things mathematics simply can't do! And one of them is to combine a collection of quantities as numerous as the points of a line segment.[4] Simple addition, for example, combines only a *finite number* of terms. The same is true of multiplication. To be honest, there *is* a modern mathematical method, which you may have encountered, called the "theory of infinite series," by means of which infinite collections of terms can, sometimes, be "added." (Such a series appears near the end of Chapter 1's *wiseacre* note.) There is also a theory of "infinite products," which extends multiplication. But the "infinity" of terms in an infinite series or product is what mathematicians call a "countable" infinity, which is much smaller than the "uncountable" infinity of points in a line segment.

There is a notion in calculus—the "definite integral"—which for a long time was thought to give the sum of uncountably many terms. To this day, in fact, mathematicians find it useful to so *interpret* it. But it was seen as *really* a sum only so long as it was loosely defined. Once a rigorous definition was formulated in the 19th century, mathematicians decided that the definite integral could be regarded as a "sum" in only an intuitive sense.

"But even if there's no mathematically rigorous way, at present, of adding enough terms, someday someone might invent one."

Even if someone already has, it wouldn't matter. All I'm trying to do, remember, is point to where your reasoning was based on intuition. When you spoke of the lengths of points as "mathematically combining" into the length of the segment, the mathematical operations you referred to simply did not exist—as far as you knew. Thus you could only have been reasoning by analogy with the operations you did know. It was, therefore, regardless of what mathematical news the future might bring, an intuitive argument.

But let me approach this issue from a different tack. Consider a rainbow—a complex phenomenon made up of water droplets suspended in the air, the sun, and an observer, all positioned relative to one another in a certain way. The rainbow is somehow the result of all these factors acting in concert, and it is difficult to assign individual responsibilities. Which factor causes the green band? Which determines the diameter? We suspect such questions are inappropriate, because if we remove any one of the contributing factors— droplets, sun, observer, or their geometrical arrangement—the entire rainbow disappears.

Perhaps a line segment is like a rainbow, only with two constituents instead of four: the points and their arrangement. Who's to say where its length comes from? The points all by themselves? But what if they were strewn randomly about the plane, like paint spatters? It seems their arrangement contributes something to the phenomenon of "length," too. But then an attempt to account for the segment's length solely in terms of the points would be doomed. Because arithmetical operations take no account of arrangement, that's precisely what you were trying to do.

"Aha!"

I know it smells like a contradiction when all those zeroes don't "add" up to 1 meter. But arithmetic is inadequate to the situation, because it ignores the geometrical aspect.

Euclid himself, by the way, probably wouldn't have had as much trouble with this as we do today. In his day the only numbers ("magnitudes") were positive; zero was not recognized. So when he said a point had no length, he *meant* exactly that, that it is improper to speak of a point as having length, as opposed to today's sense of "it has a length, but the length is zero." To him a point had no length as a water droplet has no color, and trying to deduce the length of a segment from the nonexistent lengths of its points would probably have seemed as fruitless to him as trying to deduce the colors of a rainbow, without optics, from the nonexistent colors of its droplets.

"Interesting. . . . Well. I have to admit my position *was* based on more than just logic after all. But how does it *conflict* with logic?"

Do you remember the story about the Pythagoreans and $\sqrt{2}$ (pp. 2–4)?

"Yes. The early Pythagoreans intuited that you can always find a line segment that is a 'common measure' of two other line segments, meaning its length will divide each of theirs exactly a whole number of times. It follows from this that the quotient of two lengths will always be a 'rational' number—a ratio of whole numbers—for $AB/CD = (AB/XY)/(CD/XY) = m/n$, where XY is a common measure of AB and CD and consequently m and n are whole numbers. This means that $\sqrt{2}$ is rational, because they knew how to express it as a quotient of lengths. But then one of the later Pythagoreans came up with an argument, based on logic and computation, that $\sqrt{2}$ is not a rational number."

Yes, and a beautiful argument it was, one that every mathematics student should learn by heart. Did you find it hard to follow?

"A little. I had never seen that kind of argument before. I don't mean that it was by contradiction, but that it turned on numbers being even or odd. But after a while I could follow it well enough."

What was your reaction to the Pythagoreans' original, common-sense position that two line segments always have a common measure?

"At first I didn't know what to think, the whole question was new to me. Then I decided that physically, at least, it was true. I figured that if AB and CD were sticks, for example, a carpenter could probably cut a third stick XY that laid down some whole number of times would be, for all practical purposes, the same length as AB, and that laid down some other whole number of times would be, for all practical purposes, the same length as CD. But I know geometric line segments aren't sticks, and what's close enough for practical purposes may not be mathematically exact, so in the end I had my doubts. Though their position seemed reasonable enough, I never had strong feelings that it was correct."

But according to legend, *their* feelings about its correctness were very strong. I think the legend is right, because I think their common-sense position had *not* come directly from intuition—intuitively plausible though it is—but was rather a *conclusion* they had deduced from something else.

"'Something else'?"

Something more directly intuited. Something they found utterly compelling.

"Well?"

Something about points.

"You're kidding."

No. I'm convinced that to the early Pythagoreans the diameter of a point, though extremely small, was not infinitesimal or zero. If I'm right, all along you've been an early Pythagorean!

"You mean that if points have positive diameters, we can deduce that two segments always have a common measure?"

Almost. We need *one* more assumption, but a very natural one. Tell me—back when you felt so strongly that points had positive diameters, were you picturing all those diameters as equal, or as varying from point to point?

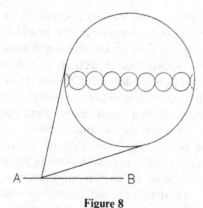

Figure 8

"... as equal. There was no reason why they should vary."

That's just what the Pythagoreans would have assumed. And because of the very same principle!—which philosophers have named "the principle of sufficient reason": variation occurs only when there is a reason sufficient to account for it.

All right, then. Every point has the same positive diameter, which we can call "d." At last we have a situation our imaginations can cope with. Mine responds with the image sketched in Figure 8, where I imagine a microscope has been aimed at a tiny portion of segment AB to show its fine structure. Is this something like what you had in mind?

"It is exactly what I had in mind."

Then consider AB/d, which for short I will call "m." Being the quotient of two positive, finite numbers, m is itself a positive, finite number. What does m represent? On the level of common sense.

"The number of points in AB."

I agree. Divide the length of AB by the length of one point, and you get the number of points. (With Figure 8 to look at, the use of simple arithmetic seems perfectly legitimate, even to an old quibbler like me!) So we see that the positive, finite number m, because it is the number of points in AB, is in fact a *whole* number.

Letting CD be any other segment, we can reason that CD is likewise made up of points with diameter d, and that $n = CD/d$ is a positive whole number, too. So a minute "line segment" consisting of only a single point is a common measure of AB and CD.

It now follows exactly as before that the quotient of two lengths is always rational, because $AB/CD = (AB/d)/(CD/d) = m/n$ and m and n are whole numbers.

"So Euclid said a point 'has no part' because $\sqrt{2}$ is irrational?"

I think so, yes, and mathematicians ever since have supported his decision. The average modern mathematician wants points to be alike as much as the

Greeks did, and this allows only the two alternatives[5] we have considered: points all have the same positive diameter d, or they all have diameter zero. As we have seen, choosing the first makes the conclusion that two segments always have a common measure inescapable, at least on the level of common sense. And while it might be possible—I don't know if it is—to choose the first alternative and yet, by logical contortions, avoid that conclusion, to do so would violate common sense to a greater extent than points without size do. So the second alternative is still the one that is chosen.

I'm glad we've had this conversation, because it has brought into sharp focus what the standard of mathematical reasoning implies[6]: that the objects with which mathematics deals not only are nonsensible—they do not physically exist—but also are at times opposed to common sense and to that degree *nonsensical*. Logic produces a "mysticism" of its own!

Euclid's Primitive Terms

There is a sense in which it doesn't matter how you visualize a point, or whether you visualize points at all.

Here are Euclid's first four Definitions:

Definition 1. A *point* is that which has no part.

Definition 2. A *line* is breadthless length.

Definition 3. The extremities of a line are points.

Definition 4. A *straight line* is a line which lies evenly with the points on itself.

Definition 1 we have discussed. The "line" of Definition 2 can be straight or curved—when Euclid intends a line to be straight, he always says "straight line," as in Definition 4. Definition 3 is an explanation of the relation between points (Definition 1) and lines (Definition 2); in it Euclid is referring to line *segments*. Sometimes he is more explicit and calls a line segment a "finite line," but even when he says only "line" he always means it to be finite. The "extremities" of a line are its ends.

As no technical term is introduced in Definition 3, it is not really a "definition" in the modern sense of that word. *Neither are Definitions 1, 2, or 4.* A "definition" (see Chapter 1, "Material Axiomatic Systems") is supposed to introduce a new technical term *solely* by means of *previously introduced* terms, except of course that unambiguous words and phrases ("the," "every," "is called," "one or more," etc.) from everyday language can be used as connective tissue. But "part" in Definition 1 is a Greek geometrical term (related to the division of figures) not previously introduced in the *Elements*; and "breadthless length" (Definition 2) and "lies evenly with" (Definition 4), while probably not technical terms—if they are, they have not been previously introduced—are by no means unambiguous phrases.

If Definitions 1, 2, and 4 are not really definitions, then the terms they introduce—"point," "line," and "straight line"—are not what we have called "defined terms." Instead they must be "primitive terms," terms which at the outset of a deductive system are introduced and explained but never precisely defined.

Though Euclid does not distinguish his primitive terms and defined terms explicitly—after all, the whole notion of "Material Axiomatic System" was abstracted from the *Elements* and similar systems only in the 19th century—he was nonetheless aware of the distinction in practice, where it counted. Recall that the only practical difference between a primitive term and a defined term is that while the latter's definition may be used to justify a step in a proof, the former's informal explanation may not. In the *Elements* Euclid treats the two types of terms in exactly this way. He freely uses the Definitions of his defined terms to justify steps, but not once in 465 demonstrations does he refer to any of Definitions 1–4 or to any of the "Definitions" (explanations) of his other primitive terms.

I began this article by saying it doesn't matter how you visualize a point. That's because Definition 1 is, logically speaking, irrelevant to the system. Since "point" is a primitive term, the only properties of "points" that may be used in a proof are those mentioned in the axioms; and Euclid's axioms (coming up) are consistent with either of the views of "point" debated in the previous article. Two people, one who agrees with Euclid that a point has diameter zero, the other who in the face of all my arguments insists that the diameter, though negligible, is positive, will nonetheless find Euclid's axioms equally acceptable, accept his deductions as well, and so in the end hold to the same geometric truths.

The logical irrelevance of Definitions 1–4 notwithstanding, it is nice to have the author of a mathematical text try to explain what he has in mind. Let us resume our examination of Euclid's Definitions from that standpoint.

As we have seen, the meaning of Definition 1 becomes clear as soon as we realize that "no part" amounts to "no size." And to apprehend Definitions 2 and 3, taken as a pair, we need only be aware that for Euclid "line" is a general term including both straight and curved lines, and that a line is taken to be finite unless Euclid explicitly says it is not. (Definition 3 does not apply to a closed curve like a circle.)

Definition 4 is curious. At this juncture most people expect Euclid to say "A *straight line* is the shortest path between two points." That famous principle does not appear, however, until the works of Archimedes,[7] some 50 years after the *Elements* was composed—and there it is an axiom, not a definition. But that's not the only reason Definition 4 is curious. What, exactly, does it mean for a line to "lie evenly with the points on itself"?

Several decades earlier, Plato had said "straight is whatever has its middle in front of both its ends" (*Parmenides* 137 E), which suggests a visual interpretation of "straight line" as line of sight—a line is "straight" if it is possible to look along it in such a way that all its points dispose themselves one in front of the other and all that can be seen is the nearest endpoint. Some scholars think that this is what Euclid had in mind in Definition 4, and that what prevented him from simply saying so was his desire for rigor—if mathematics was to be kept insulated from its down-to-earth origins, it wouldn't do to define a fundamental concept like "straight line" in terms of a physical phenomenon like human vision.

Other scholars, skeptical of Platonic influence, think that a stretched fiber was actually Euclid's image of "straight line"—though agreeing that what prevented him from simply saying so was his struggle to avoid reference to physical objects.

I personally have no idea what Euclid meant by Definition 4. Another of his books (the *Optics*) makes it clear, however, that he did consider a line of sight (the path of a ray of light) to be at least a *physical manifestation* of a straight line.

Definition 5. A *surface* is that which has length and breadth only.

Definition 6. The extremities of a surface are lines.

Definition 7. A *plane surface* is a surface which lies evenly with the straight lines of itself.

Definitions 5, 6, and 7 are the two-dimensional analogs of Definitions 2, 3, and 4, and like the latter are not definitions but explanations of primitive terms. "Surface" (Definition 5) is a general term for all surfaces, be they flat, curved, undulating, whatever. A surface is usually taken to be finite in extent; so if it is not closed like a sphere it has a boundary, which Definition 6 explains is made up of lines.

Definition 7 is just as obscure as Definition 4, but this time only the visual interpretation is easy to understand: a "plane surface" is a surface that edge-on appears to be a straight line. In practice Euclid shortens "plane surface" to "plane."

Though many of the theorems in Book I of the *Elements* are true for figures arbitrarily oriented in 3-dimensional space, others require the setting to be a single plane. Our approach will be to view them *all* as set in a single (infinite) plane, which we will refer to as simply "the plane." (As usual Euclid is mute, but as he does not explicitly begin 3-dimensional geometry until Book XI, it appears his view was the same.) This will keep our drawings simple and harmonize with the reader's geometric experience, which is probably confined to plane geometry anyway. As a result we will have little occasion to use the primitive terms "surface" and "plane surface."

As the "Definitions" after Definition 7 really *are* definitions and so there are no more primitive terms, we see that Book I involves a total of five primitive terms, only three of which will concern us: "point," "line," and "straight line."

Euclid's Defined Terms (Part 1)

Definition 8. A *plane angle* is the inclination to one another of two lines in a plane which meet one another and do not lie in a straight line.

"Hold on!" I imagine someone cry. "You said the Definitions after Definition 7 are real definitions. But what about the word 'inclination'? It's not a previously introduced technical term. It's more than 'connective tissue'. And it's certainly not unambiguous."

True, true. The same case can be made, though probably not as strongly, against the word "meet," and perhaps other parts of Definition 8 as well. As it stands it is not a real definition. But unlike Definition 7, it can *become* a real definition if it is rewritten. For example

A *plane angle* consists of two lines in a plane which meet one another and do not lie in a straight line

circumvents any objection that might be made to "inclination," and more complicated revisions can be made to meet other objections.

Remember that Euclid had no reason to fashion his Definitions in strict adherence to a distinction (between primitive and defined terms) that would not be made explicitly for 2,200 years. Nonetheless, the power of the unmade distinction can be felt. Definitions 1–7 and 8–23 have an unmistakable difference in flavor, which derives from the fact that the former *cannot* be reworked into real definitions while the latter *can*. It is in this sense we will call the terms of Definitions 8–23 Euclid's "defined terms," even though we will not take the time to purge those Definitions of foreign references.

Note that the sides of a "plane angle" are merely "lines"—not "straight lines"—and so can be curved. Euclid probably allowed angles with curved sides only in deference to their then common recognition as legitimate angles. Only one theorem in the *Elements* (Book III, Theorem 16) involves such an angle (see Figure 9); the theorem is never used, and even in that case one side of the angle is straight.

Euclid's term for a normal plane angle with two straight sides is translated "plane *rectilinear* angle," from the Late Latin *rectus*, straight + *linea*, line.

Definition 9. And when the lines containing the angle are straight, the angle is called *rectilinear*.

As we will never have occasion to discuss angles with curved sides I will refer to "plane rectilinear angles" as simply "angles," a condensation Euclid uses himself.

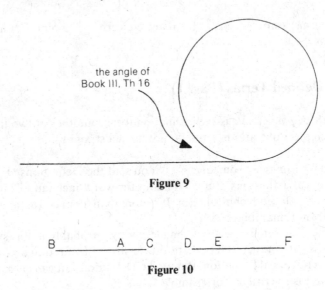

the angle of
Book III, Th 16

Figure 9

B_____A___C____D__E_____F

Figure 10

Since Definition 8 calls the sides of an angle "*two* lines," meaning they are wholly distinct, and specifies that they "do not lie in a straight line," objects like *ABC* and *DEF* in Figure 10 are excluded as "angles." For Euclid every angle is—to use modern terminology—greater than 0° and less than 180°.

Definition 10. When a straight line set up on a straight line makes the adjacent angles equal to one another, each of the equal angles is *right* , and the straight line standing on the other is called a *perpendicular* to that on which it stands.

This may not be how you would first conceive of a right angle. See Figure 11. If a straight line *CD* "set up on" a straight line *AB* makes the "adjacent" angles 1 and 2 equal, then angles 1 and 2 are called *right* angles and *CD* is called a *perpendicular* to *AB*.

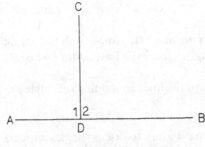

Figure 11

"Sufficient for Each Day Is the Rigor Thereof" [8]

Scholars scrutinizing the *Elements* have brought to light a great many things that Euclid assumed but did not bother to write down. As a rule it will suit our purpose to let such things lie undisturbed in the comfy shadows of common sense. Each disclosure complicates our study; and reading the *Elements* we will very naturally come to the same tacit understandings as the author, anyway. As our itinerary includes a tour through a non-Euclidean geometry, however, there are certain unstated assumptions we need to make explicit. This we will do as we encounter them.

Indirectly, Definitions 11 and 12 involve a tacit understanding. Though it is *not* one of the ones we need to expose, I will expose it to show you the *kind* of thing we will henceforth be absorbing uncritically.

Definition 11. An *obtuse* angle is an angle greater than a right angle.

Definition 12. An *acute* angle is an angle less than a right angle.

Presupposed here is that we know, given two unequal angles, how to determine which is "greater."

It would be worthless to physically measure the angles, whether by eye or an instrument, because we couldn't *use* our finding—"by measurement" is not one of the reasons allowed (p. 12) to justify a step in a proof. Remember that mathematics keeps itself at a distance from the physical world, of which geometric diagrams are a part. "The geometer bases no conclusion on the particular line which he has drawn being that which he has described, but [he refers to] what is *illustrated* by the figures," said Aristotle. There is the additional factor that even if we went ahead and measured the angles anyway, we couldn't *trust* our finding. Geometers, realizing it is impossible for a drawing to be mathematically exact, and agreeing with Aristotle that their drawings are only supposed to be visual aids in the first place, have always tended to put into them only what people need in order to follow the argument; they don't worry if an angle, say, is drawn a little too big or too small. "Geometry," runs a maxim modern geometers are fond of repeating, "is the art of applying good reasoning to bad diagrams."

So what does Euclid have in mind? Immediately following the Definitions are Euclid's axioms, the last of which gives a clue. "The whole," it says, "is greater than the part." One angle will be "greater" than another, Euclid is telling us, if it is a "whole" of which the other is a "part." But to what relationship between angles do the terms "whole" and "part" refer? Euclid doesn't say, so in a way we're back where we started, but his language does suggest something like Figure 12, where the "whole" $\angle ABC$ is made up of the "parts" $\angle ABD$ and $\angle DBC$. Though at this point it is by no means clear that this relationship between $\angle ABC$ and (say) $\angle DBC$ really is the kind of rela-

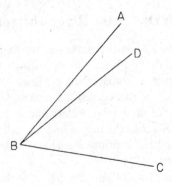

Figure 12

tionship two angles would have to have for Euclid to consider them "whole" and "part"—and so for ∠ABC to be "greater" than ∠DBC—perusal of Euclid's proofs reveals that it is. Whenever he asserts, without appeal to a theorem, that one angle is greater than another, the angles always share the vertex and one side, as in our figure, and the "greater" is always the one whose sides flank a side of the other. Apparently Euclid is willing to accept the testimony of a diagram *to an extent*: Figure 12 could not be so badly drawn, he seems to feel, that BD has been placed on the wrong side of BA.

We conclude that Euclid's understanding of what it means for an angle to be "greater" than another can be stated as follows.

Given an angle ABC and a straight line BD, with the points A and D on the same side of BC (Figure 12), we say that angle ABC is *greater* than angle DBC if BD is between BA and BC.

The meaning of "less" than now follows immediately: one angle is *less* than another if in the above sense the latter is greater than the former.

I insisted that A and D be on the same side of BC (Euclid would have, too) to avoid situations like Figure 13. There BD *is* in some sense "between" BA

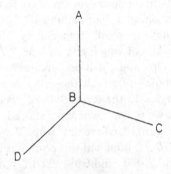

Figure 13

and BC but Euclid wouldn't have called $\angle ABC$ and $\angle DBC$ "whole" and "part."

Definitions 11 and 12 involve another assumption as well, which so far has not been stated. It is not really a "tacit" assumption, however, because Euclid will soon state it quite baldly as another of his axioms. We will discuss it when we get to that axiom; but in the meantime, can you detect what this other assumption is?

In 1800 most mathematicians revered the *Elements*, which they regarded as the supreme example of airtight deductive presentation. By 1900, due to crises in the foundations of several branches of mathematics, in particular to the crisis in geometry precipitated by the invention of non-Euclidean geometry, most mathematicians were contemptuous of the old masterpiece and regarded it as a logical sieve. It had been examined more ruthlessly than ever before and found to be shot through with intuitive notions the Greeks and their successors had overlooked.

An example is the word "between," which appears in the *Elements* and which we have just used (p. 36) to make explicit Euclid's understanding of what it means for an angle to be "greater" than another. (Stating one tacit assumption involved making another!) Admittedly the notion conveyed by "between" seems clear enough. In Figure 12, BD is "between" BA and BC—what could be less ambiguous than that? But if we examine the clause of our statement in which the word is contained—"we say that angle ABC is *greater* than angle DBC if BD is between BA and BC"—we see that, unlike the words and phrases that are clear-cut connectives—"we say that," "is," "if," "and"—"between" carries a meaning which is unquestionably *geometrical*, making it something that, strictly, ought to be a technical term. (We ought not accept the testimony of a diagram *at all*.) As we have neither defined the word nor included it as a primitive term, it stands exposed as intuitive. "On the same side," also in the *Elements* and also in our statement, is another example of an inconspicuous expression whose meaning is geometrical but whose only foundation is intuitive.

And so there was undertaken a second great purging of intuition from geometry, and a rebuilding of the subject from the ground up. The best-known reformulation is *Foundations of Geometry* by the mathematical giant David Hilbert (1862–1943), first published in 1899 (a 10th edition appeared in 1968). Hilbert's primitive terms are *point*, *straight line*, *plane*, *on* (as in "point C is on straight line AB"), *between*, and *congruent* (corresponding to Euclid's "equal"). His axioms number fifteen (Euclid has ten), and his theorems include such intuitively obvious statements as

Theorem 4. *Of any three points A, B, C on a straight line there is always one that lies between the other two.*

Thus geometry became "rigorous" beyond the Greeks' wildest dreams. By today's hard-line standards the *Elements* is indeed a logical sieve. Nevertheless

I have organized this book around the *Elements* rather than *Foundations of Geometry*, as I feel it will carry us more quickly and pleasantly to the real issues. Caulking every little logical crevice is tedious work, and unless you have extensive mathematical experience you would likely find it confusing as well. More importantly, it is Euclid's text, not Hilbert's, that has been the scientific paradigm for most of scientific history; and it is Euclid's, not Hilbert's, to which the inventors of non-Euclidean geometry were responding.

We'll plug the major holes in the *Elements*, root out the intuitive components that are likely to cause trouble when we get to non-Euclidean geometry, and let it go at that. Harmless intuitive references, like those just discussed, we will allow to remain; we won't bother to construct the involved logical substructure with which, as Hilbert has shown, they can be supported.

Euclid's Defined Terms (Part 2)

We won't need Definitions 13 or 14, so I will omit them.

Definition 15. A *circle* is a plane figure contained by one line such that all the straight lines falling upon it [the *radii*] from one point among those lying within [it] are equal to one another;

Definition 16. And the point is called the *center* of the circle.

The "one line" of Definition 15 is the circle itself. The "straight lines falling upon it" from the center (Definition 16) that "are equal to one another" are what we would call its "radii," so I have interpolated that term. Circles are the only nonstraight lines that occur in the *Elements*.

We won't need Definitions 17 or 18, so I will omit them also.

Definition 19. *Rectilinear* figures are those which are contained by straight lines, *trilateral* figures [*triangles*] being those contained by three, *quadrilateral* those contained by four, and *multilateral* those contained by more than four straight lines.

Definition 20. Of trilateral figures, an *equilateral triangle* is that which has its three sides equal, an *isosceles triangle* that which has two of its sides . . . equal, and a *scalene triangle* that which has its three sides unequal.

The original reads "an *isosceles triangle* that which has two of its sides alone equal." I have deleted "alone" to bring Euclid's definition into line with modern usage, according to which an equilateral triangle is an isosceles triangle (of a special kind).

Definition 21. Further, of trilateral figures, a *right-angled triangle* is that which has a right angle, an *obtuse-angled triangle* that which has an obtuse angle, and an *acute-angled triangle* that which has its three angles acute.

Here Euclid is telegraphing the fact, to be established in Theorem 17, that a triangle can have at most one nonacute angle.

Definition 22. Of quadrilateral figures, a *square* is that which is both equilateral and right-angled . . . .

In other words, a four-sided figure with four equal sides and four right angles.

Finally we come to

Definition 23. *Parallel* straight lines are straight lines which, being in the same plane and being produced indefinitely in both directions, do not meet one another in either direction.

Had Euclid's standard straight line been infinite instead of finite, he would have said simply

Parallel straight lines are straight lines which, being in the same plane, do not meet one another in either direction.

Since his usual "straight line" was a segment, however, he had to add the clause about "producing" (extending) them.

Euclid's Axioms

Euclid states 10 axioms, though there are others he didn't bother to write down, some of which we will uncover on our way through Book I. Of the 10 explicit axioms, he calls the first five "Postulates" and the last five "Common Notions." His distinction, not made by modern geometers, seems to have been that whereas the Postulates are specifically geometrical assumptions, the Common Notions are basic to other sciences as well.

Postulate 1. [It is possible] to draw a straight line from any point to any point.

Postulate 2. [It is possible] to produce a finite straight line continuously in a straight line.

Postulate 3. [It is possible] to describe a circle with any center and distance [radius].

The first three postulates empower us to construct things: to connect two points by a straight line whenever we like, to prolong ("produce") any finite straight line, and to draw ("describe") circles, of any size, centered anywhere. Postulates 1 and 2 are called the "straightedge postulates" because they metaphorically give us unrestricted use of a pencil and unmarked straightedge. Similarly Postulate 3 is called the "compass postulate."

Permitting these constructions is Euclid's way of stipulating the existence of the objects constructed. Every branch of mathematics needs a clear-cut standard whereby proposed objects of study inconsistent with its axioms can be recognized as lying outside that branch's legitimate purview. Such objects are declared "not to exist," and the standard that bars them is called the "existence criterion." (1/0, for example, does not meet the existence criterion of the standard number system, because its admission would lead to absurdities like $1 = 2$.[9] Thus we are told in school that division by 0 "is impossible," or that 1/0 "is undefined" or "does not exist.") Euclid's existence criterion is constructability with compass and straightedge.

Euclid clearly understood the straight line of Postulate 1 and the circle of Postulate 3 to be unique—that joining two points there exist *only one* straight line, and that there exist *only one* circle having a given center and radius. It is also clear from his use of Postulate 2 that he understood it to allow producing a finite straight line, from either end, by as *much* as we want. To make these understandings explicit we shall revise Postulates 1–3 as follows.

Postulate 1. It is possible to draw one and only one straight line from any point to any point.

Postulate 2. From each end of a finite straight line it is possible to produce it continuously in a straight line by an amount greater than any assigned length.

Postulate 3. It is possible to describe one and only one circle with any center and radius.

The next postulate strikes many people as redundant.

Postulate 4. All right angles are equal to one another.

"Of *course* all right angles are equal," I imagine someone saying. "They're all equal to 90°!" We are so used to measuring angles in "degrees" that it's easy to overlook the fact that Euclid has made no mention of that term (he never does). Nor has he given us a protractor among our metaphorical tools. The symbolic "90°" is, admittedly, a handy shorthand for "right angle," and we will introduce it shortly for that purpose; but even once this is done "right angle" will remain logically prior. We will know that "90°" always represents the same angle-size because of Postulate 4, not the other way around.

While Definition 10 does say right angles come in equal pairs, it does not

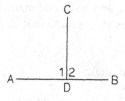

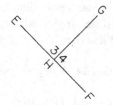

Figure 14

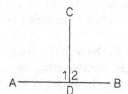

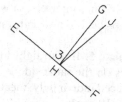

Figure 15

compel us to believe two right angles in one part of the plane are equal to two others somewhere else. Suppose the two drawings of Figure 14 are billions of miles apart. If $\angle 1 = \angle 2$ and $\angle 3 = \angle 4$, then by Definition 10 each of angles 1, 2, 3, and 4 is properly called a "right" angle. But is $\angle 1 = \angle 3$? It's true that calling both $\angle 1$ and $\angle 3$ "right" angles *suggests* they are equal; but on that ground the angles we call "acute" should all be equal. Couldn't it be that the plane's character evolves over the vast distance between $\angle 1$ and $\angle 3$? In fact wouldn't it be remarkable if the plane were uniform over its entire un-limited extend? But that is just what we imply when we say that, necessarily, $\angle 1 = \angle 3$.

My point is that the truth of Postulate 4 is not obvious. It tells us something we didn't know before—that the plane *is* uniform, at least to the extent that right angles are the same no matter where they are.

Only now do "acute" and "obtuse" take on their usual significance. These terms are so familiar that when reading their Definitions (11 and 12) you may not have detected any lack, even with my talk of a hidden assumption (p. 37). Come back with me for a moment to the time before Postulate 4. In Figure 15 let $\angle 1 = \angle 2$ and $\angle 3 = \angle GHF$, so all four angles are right by Definition 10. $\angle JHF$ is less than $\angle GHF$, so $\angle JHF$ is acute by Definition 11. The familiarity of the situation can lull us into thinking that $\angle JHF$ is automatically less than $\angle 2$. But without Postulate 4, that does not follow! Without Postulate 4 $\angle GHF$ could be, for all we know, a lot bigger than $\angle 2$, in fact so much bigger than even the smaller $\angle JHF$ would also be bigger than $\angle 2$. In view of this possibility our application of the term "acute" to $\angle JHF$ would have signified exactly nothing about its absolute size. *With* Postulate 4, however, we know that "right angle" is a fixed and universal standard to which other angles can be compared, enabling us to conclude that $\angle JHF$ is indeed less than $\angle 2$, as desired.

Definition 24. A *degree* is one-ninetieth of a right angle.

Adding this Definition to Euclid's list enables us to write "90°" for "right angle," and "180°" for Euclid's cumbersome "two right angles."

The purpose of Postulate 5 is to enable us to conclude certain pairs of straight lines intersect, in which case it goes on to say where the intersection takes place. It is noticeably more complicated than Euclid's other nine axioms.

Postulate 5. If a straight line falling on two straight lines make the interior angles on the same side less than two right angles, the two straight lines, if produced indefinitely, meet on that side on which are the angles less than the two right angles.

A drawing will help. In Figure 16, *EF* is a straight line "falling" on two straight lines *AB* and *CD*. (Modern textbooks would call *EF* a "transversal.") There are two pairs of "interior angles on the same side": angles 1 and 2, and angles 3 and 4. Postulate 5 says that if either pair adds up to less than 180° then *AB* and *CD*, if extended far enough, will intersect on the same side of *EF*. Specifically, if $\angle 1 + \angle 2 < 180°$ then *AB* and *CD* will meet to the right of EF (in my drawing), and if $\angle 3 + \angle 4 < 180°$ they will meet to the left. Before Euclid makes use of Postulate 5 he will prove that it is impossible for $\angle 1 + \angle 2$ and $\angle 3 + \angle 4$ to *both* be less than 180°.

Note that in treating 180° ("two right angles") as a fixed quantity, Postulate 5 depends for its significance on Postulate 4.

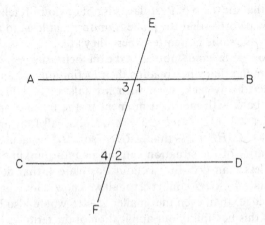

Figure 16

Common Notions

1. Things which are equal to the same thing are also equal to one another.
2. If equals be added to equals, the wholes are equal.
3. If equals be subtracted from equals, the remainders are equal.
4. Things which coincide with one another are equal to one another.
5. The whole is greater than the part.

You may recognize Common Notion 5 as the axiom we referred to on page 35.

The Common Notions are Euclid's last explicit axioms. They are general rules for reasoning with quantities—or, to use the Greek term, "magnitudes." In plane geometry these magnitudes are of three kinds: lengths of finite lines, angle sizes, and areas.

All but Common Notion 4 can be expressed algebraically, where "a," "b," "c," and "d" represent any magnitudes of the same kind.

Common Notion 1. If $a = b$ and $c = b$ then $a = c$.

Common Notion 2. If $a = b$ and $c = d$ then $a + c = b + d$.

Common Notion 3. If $a = b$ and $c = d$ then $a - c = b - d$.

Common Notion 5. $a + b > a$.

Common Notion 4 cannot be expressed algebraically because, unlike the other Common Notions, it involves a notion that is not quantitative: coincidence. This is a specifically geometric notion, which it connects to the quantitative one of equality. Taking it at face value, Common Notion 4 is what justifies our saying (Figure 17) that side AD of triangle ABD is equal to side AD of triangle ACD. But it seems Euclid had another, even more geometric, use for Common Notion 4, namely to justify a controversial procedure he uses reluctantly on two occasions (Theorems 4 and 8) and which Hilbert, in his reformulation of Euclidean geometry, came to reject altogether—the superposition of figures. I'll tell you the whole story when we come to Theorem 4.

Unfortunately there are simple quantitative deductions you or I would

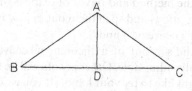

Figure 17

naturally make, and Euclid in fact does make, which are unjustified by any Common Notion. If we knew for instance that $\angle 1 = \angle 2$ and $\angle 3 < \angle 4$, we would naturally conclude that $\angle 1 + \angle 3 < \angle 2 + \angle 4$. But no Common Notion says that when equals are added to *unequals*, the sums are unequal[10] (in the same way). We could perhaps distill some sort of justification out of Common Notions 2 and 5, but the labor of doing so would be considerable, and would have to be repeated several times before we had established all the principles of quantitative reasoning Euclid uses. Another possibility, since we are, after all, in the business of stating baldly some of the things that are only implicit in the *Elements*, would be to add a series of *new* "Common Notions," covering every type of quantitative deduction we might have occasion to make; but such a series would be discouragingly long. As a compromise I will add a single "Common Notion" that recognizes, in general terms, the legitimacy of treating geometric magnitudes like numbers, with the understanding that whenever we apply it we will state the specific numerical principle we have in mind.

Common Notion 6. Equations and inequalities involving geometric magnitudes of the same kind obey the same laws as equations and inequalities involving positive numbers.

Thus, to return to our example, if we knew that $\angle 1 = \angle 2$ and $\angle 3 < \angle 4$, we would write the following as the reason for our conclusion that $\angle 1 + \angle 3 < \angle 2 + \angle 4$:

C.N. 6 (If $a = b$ and $c < d$ then $a + c < b + d$.)

Technically, Common Notion 6 includes all the other Common Notions except Common Notion 4, making them redundant. However, I will continue to use these others whenever they apply, reserving Common Notion 6 for situations in which they don't.

Theorems Proven Without Postulate 5

Euclid seems to have put off using Postulate 5 as long as he could. At any rate, the first 28 theorems of Book I are proven without it, as is Theorem 31 (Book I has 48 theorems in all). We will examine most of the Postulate 5-independent theorems in this section. Along the way I hope to accomplish three things: reacquaint you with the method and flavor of plane geometry; uncover more of Euclid's tacit assumptions and spell them out in new Postulates; and patch up the difficulty with Theorems 4 and 8.

I will present Euclid's proof of a theorem whenever it is problematic, instructive, or especially admirable. Other proofs will be left for you to invent, on the theory you might like to try your hand. (If you wouldn't, just accept the theorems, or look up their proofs in the *Elements*.)

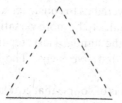

Figure 18

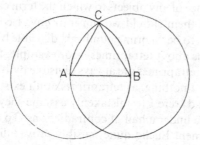

Figure 19

Theorem 1. *On a given finite straight line [it is possible] to construct an equilateral triangle.* (Figure 18.)

Proof.

1. Let "*AB*" be the given finite straight line (see   hypothesis
   Figure 19).
2. Draw a circle with center *A* and radius *AB*.   Post 3
3. Draw a circle with center *B* and radius *AB*.   Post 3
4. Let "*C*" be a point where the circles intersect.
5. Draw *AC* and *BC*.   Post 1
6. *AC* = *AB*   Def "radius"
7. *BC* = *AB*   Def "radius"
8. *AC* = *BC*   6, 7, C.N. 1
9. *ABC* is an equilateral triangle.   6, 7, 8, Def "equilateral
                                            triangle"
10. Therefore, on a given finite straight line it is   generalization
    possible to construct an equilateral triangle.

A note on format. Euclid does *not* arrange his demonstrations in two columns. I have recast them because it makes them easier to analyze. Also, I will refer to Definitions by the term involved, but to Postulates, Common Notions, and Theorems by number. (As the Postulates and Theorems accumulate and it becomes difficult to recall them by number, refer to the "Index to Euclidean Geometry" on pp. 97–99.)

Postulates 1 and 3 empowered us to construct straight lines and circles. This theorem shows those Postulates to be more versatile than they appear if we use them together. Apart from the standard last step the proof is in two parts. The triangle is erected in the first five steps. The next four verify that it is equilateral, as advertised.

The import of Theorem 1 is of course that equilateral triangles exist. From a logical point of view it is essential to verify this if we are to have equilateral triangles available for future use (e.g., in the proof of Theorem 2). Just defining a term, as we did "equilateral triangle" in Definition 20, doesn't guarantee the existence of any objects to which the term can be applied; it only tells us what to call them should we run into any. I could define a "tetra-prime," for instance, to be "a prime number[11] divisible by 4." I could even go on to prove theorems about tetraprimes—for example, that no tetraprime is odd, that squares of tetraprimes are always divisible by 16, etc. But I would be literally talking about nothing, as tetraprimes don't exist. (No prime number is divisible by 4.) And were I to mistakenly assume they *did* exist, my arithmetic would collapse under a hail of contradictions. To avoid the possibility of either embarrassment Euclid quite sensibly proves the existence of whatever geometric objects he plans to discuss.

There is a bit of a problem, though, with step 4. Things can be named as soon as we know they exist, but how do we know point C exists? I grant that the circles interpenetrate, and in the drawing it certainly *looks* as if they share a point, in fact two points. But "see drawing" is not a mathematically acceptable reason. A drawing is not part of a proof, only an aid to following it. And what if I had drawn a different picture? Maybe the points on the circles have spaces between them, as in Figure 20. Interpenetrating necklaces don't share a bead.

I'm not seriously maintaining that circles might be like necklaces. The question is not, "Do interpenetrating circles intersect?" Of course they intersect. You believe they do. I believe they do. Euclid certainly believed they do. And this is *our* system. The question is rather, "On what basis?" What do we give as a reason for step 4? We have no Definition or axiom that says point C must exist.

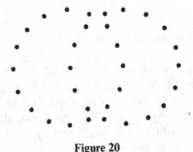

Figure 20

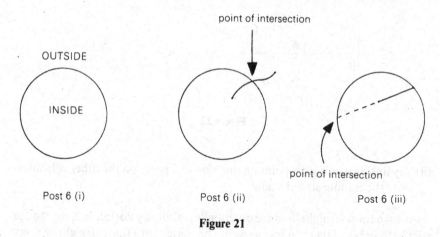

Figure 21

The easy solution is to simply set down a new Postulate as an official record of what we already believe.

Postulate 6.[12]

(i) A circle (or triangle) separates the points of the plane not on the circle (triangle) into two regions called its *outside* and *inside* (see Figure 21);

(ii) any line drawn from a point outside to a point inside intersects the circle (triangle); and

(iii) any straight line drawn from a point on the circle (triangle) to a point inside will, if produced indefinitely beyond the point inside, intersect the circle (triangle) exactly one more time.

The three parts of the postulate are illustrated here for a circle. Part (i) sets the stage for (ii). Part (iii) is an extension of (ii) which we will need later on. Part (ii) is what we need now to tighten up the proof of Theorem 1, as follows:

4. The circles intersect at "*C*".   Post 6 (ii)

In applying Postulate 6 (ii) we are taking one circle as the "circle" and an arc of the other circle as the "line" drawn from a point outside to a point inside.

Postulate 6 is called a "continuity" axiom, because its overall effect is to assure us that circles and triangles are continuous figures, with no gaps between their points. Later we will need another continuity axiom; as it is the analog for straight lines of Postulate 6, we may as well state it now.

Postulate 7.

(i) A straight line extending infinitely far in both directions separates the points of the plane not on it into two regions called its *sides* (see Figure 22); and

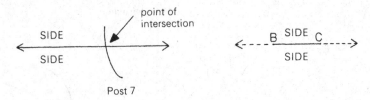

Post 7

Figure 22

(ii) any line drawn from a point on one side to a point on the other side intersects the infinite straight line.

As even finite straight lines are potentially infinite by Postulate 2, we can use part (i) to speak of their "sides" as well. The "sides" of a finite straight line, say *BC*, are the sides of the infinite straight line consisting of all points through which *BC* could be produced.

Theorems 2 and 3 ... are mysterious to all who do not see that Postulate 3 does not ask for *every use of the compasses*.
                                   —Augustus De Morgan,[13] 1849 (in Heath's *Euclid*, p. 246)

There's a subtlety to Postulate 3 we should take up before going on, or Theorems 2 and 3 won't make sense. It is this: the metaphorical compass we are given to use in Postulate 3 won't stay open. When either leg is lifted from the plane it falls shut. It can't be used as a pair of dividers to transfer distances.

That is—dropping the metaphor—Postulate 3 can be used to draw a circle with a given point as center and a given straight line as radius, *only* if the point is an endpoint of the straight line. The proposed center and radius must come already attached. In Figure 23 for example we can draw only two circles with radius *BC*—one with center *B*, the other with center *C*. The Postulate does not permit us to draw one with center *A*. Such a circle would have radius *equal* to *BC*. In Greek mathematics there's a difference between saying "the radius is *BC*" and "the radius is equal to *BC*." Nowadays the distinction is lost because we tend to think of a radius as a number, a free-floating entity. According to Definition 15, however, a "radius" is one of the equal straight lines emanating from the center of a circle, so strictly speaking it is a *geometric* object and as

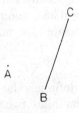

Figure 23

such is tied to a particular place. Postulate 3 gives us the power to draw a circle one of whose radii *is BC*, so there are the two possibilities, but it doesn't give the power to draw a circle all of whose radii are merely *equal* to *BC*.

Theorems 2 and 3 say that even though the postulated compass keeps collapsing, it can still be used, albeit in a roundabout way, to do everything a normal compass can. Then why on earth didn't Euclid simply postulate a normal compass in the first place? Probably out of pride. Among mathematicians it is "inelegant" to assume more than one has to.

Specifically Theorem 2 tells us how to attach, to a given point, a segment equal to a given segment located elsewhere. It does not, however, allow us to control the direction in which the constructed segment will be drawn.

Theorem 2. [*It is possible*] *to place at a given point* [*as an extremity*] *a straight line equal to a given straight line.* (Figure 24.)

Proof.
1. Let "*A*" be the given point and "*BC*" the given straight line (see Figure 25). — hypothesis
2. Draw *AB*. — Post 1
3. On *AB* draw an equilateral triangle *AB*"*D*". — Th 1
4. Draw a circle with center *B* and radius *BC*. — Post 3
5. Produce *DB* to a point "*E*" outside the circle. — Post 2
6. *BE* intersects the circle in a point "*F*". — Post 6 (ii)

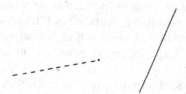

Figure 24

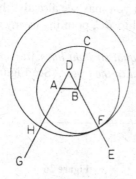

Figure 25

7. Draw a circle with center D and radius DF.   Post 3

8. Produce DA to a point "G" outside this second   Post 2
   circle.

9. DG intersects the second circle in a point "H".   Post 6 (ii)

10. $DH = DF$ Def "radius"

11. $DA = DB$ 3, Def "equilateral
                                                       triangle"

12. $AH = BF$ 10, 11, C.N. 3

13. $BC = BF$ Def "radius"

14. $AH = BC$ 12, 13, C. N. 1

15. AH has the given point A as an extremity and is   1, 14
    equal to the given straight line BC.

16. Therefore, it is possible to place at a given point   generalization
    as an extremity a straight line equal to a given
    straight line.

Henceforth I will omit the traditional last step in which the theorem is obtained by generalization.

Like that of Theorem 1 the proof is in two parts. Steps 1–9 provide the blueprint for the construction, then steps 10–15 check that the outcome AH meets specifications.

Note that in step 5, Postulate 2 is invoked to prolong DB by an amount greater than the radius BC of the circle. Since BC could have been any length—it was the arbitrary straight line given at the outset—apparently we can produce finite straight lines by as much as we want. They are potentially infinite. This is why I modified Postulate 2 as I did (p. 40).

In the construction AH is an extension of DA, whose direction is fixed by the given positions of A and BC, so we have no control over the direction in which AH is drawn.[14]

A little while ago I said, referring to Figure 23, that Postulate 3 does not allow us to draw a circle with center A and radius merely *equal* to BC. Now, with Theorem 2, we can: use Theorem 2 to attach to A a straight line $AH = BC$, then use Postulate 3 to draw a circle with center A and radius AH. As Theorem 3 will show, and you may see already, this is the key to drawing a straight line from A, equal to BC, pointing in any direction we want.

Theorem 3. *Given two unequal straight lines, [it is possible] to cut off from the greater a straight line equal to the lesser.* (See Figure 26.)

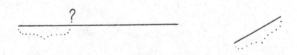

Figure 26

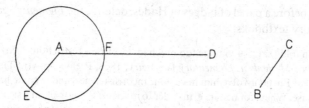

Figure 27

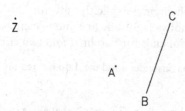

Figure 28

Proof.

1. Let "*AD*" and "*BC*" be the straight lines with,   hypothesis
   say, *AD* greater than *BC* (see Figure 27).
2. From *A* draw a finite straight line *A*"*E*" = *BC*.   Th 2
3. Draw a circle with center *A* and radius *AE*.   Post 3
4. *AD* > *AE*   1, 2, C.N. 6 (If $a > b$
   and $c = b$ then $a > c$.)
5. The circle intersects *AD* in a point "*F*."   Post 6 (ii)
6. *AE* = *AF*   Def "radius"
7. *AF* = *BC*   2, 6, C.N. 1
8. *AF* has been cut off from *AD* and is equal to *BC*.   5, 7

I introduced Theorem 3 saying that, given a point *A* and a straight line *BC*, it would enable us to draw a straight line from *A* equal to *BC* and pointing in any direction we want. Suppose the direction we want is toward point *Z* (see Figure 28). Draw *AZ*; and if *AZ* is not longer than *BC*, produce it by Postulate 2 until it is. Then use Theorem 3 to cut off *AF* equal to *BC*.

We will use Theorem 3 frequently. Concerning a possible misunderstanding as to its application, and that of Euclid's construction theorems in general, I'd like to quote an exchange from Charles L. Dodgson's (Lewis Carroll's) *Euclid and His Modern Rivals*,* an unintentionally hilarious play in which the ghost

---

* Reprinted from Charles L. Dodgson: *Euclid and His Modern Rivals*, 1879; Dover Publications, Inc., New York, 1973, with permission.

of Euclid, before a panel of judges in Hades, defends the *Elements* against rival 19th-century textbooks.

Minos. I am told that you indulge too much in "arbitrary restrictions." Mr. Reynolds says (*Modern Methods in Elementary Geometry*, 1868, Preface, p. vi) "The arbitrary restrictions of Euclid involve him in various inconsistencies, and exclude his constructions from use. When, for instance, in order to mark off a length upon a straight line, he requires us to describe five circles, an equilateral triangle, one straight line of limited, and two of unlimited length, he condemns his system to a divorce from practice at once and from sound reason."

(While the proof of Theorem 3 explicitly calls for only one circle, it invokes Theorem 2, whose proof calls for one line and two circles and in turn invokes Theorem 1, whose proof calls for two lines and two circles.)

Euclid. Mr. Reynolds has misunderstood me: I do not require all that construction in Theorem 3.

I merely prove, once and for all, in Theorem 2, that a line *can* be drawn, from a given point, and equal to a given line, by the original machinery alone, and *without* transferring distances. After that, my reader is welcome to transfer a distance by any method that comes handy . . . and of course he may now transfer his compasses to a new center. And this is all I expect him to do in Theorem 3.

Minos. Then you *don't* expect these five circles etc. to be drawn whenever we have to cut off, from one line, a part equal to another?

Euclid. Pas si bête, mon ami [My friend, it's not *that* stupid].

—Act IV, Scene 2

Theorem 4. *If two triangles have the two sides equal to two sides respectively, and have the angles contained by the equal straight lines equal, they will also have the base equal to the base, the triangle will be equal to the triangle, and the remaining angles will be equal to the remaining angles respectively, namely those which the equal sides subtend.*

This is the Side–Angle–Side (SAS) criterion for (what we now call) "congruence" of triangles. Refer to Figure 29. "If two triangles have the two sides

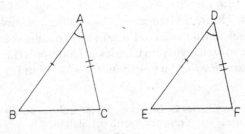

Figure 29

equal to two sides respectively" means that, for example, $AB = DE$ and $AC = DF$, in which case "and have the angles contained by the equal straight lines equal" would mean $\angle A = \angle D$. The conclusion is in three parts:

(1) "They will also have the base equal to the base", i.e., $BC = EF$. Euclid often calls the bottom of a figure the "base," but as the correctness of the theorem would not be affected by looking at the drawing, say, upside-down, the term has no mathematical meaning and in this case simply refers to the third side of each triangle.

(2) "The triangle will be equal to the triangle", i.e., the triangles have equal *areas*.

(3) "And the remaining angles will be equal to the remaining angles respectively, namely those which the equal sides subtend," i.e., $\angle C = \angle F$ and $\angle B = \angle E$. The Greek phrase here translated "subtend" means "stretch under"–"subtend" is the Latin equivalent—so when Euclid says a side "subtends" an angle he refers to the side *opposite* that angle. By hypothesis AB and DE are equal sides and the angles they subtend are $\angle C$ and $\angle F$, so (3) asserts that $\angle C = \angle F$, and similarly that $\angle B = \angle E$.

I hate to tamper with Euclid's enunciation, but we should not allow so important a theorem to lie hidden behind antique phraseology. (It turns out we will have to discard his proof, anyway.) Let's introduce the modern term "congruent."

Definition 25. Two triangles are *congruent* if the three angles of one are equal respectively to the three angles of the other and the sides subtending equal angles are equal.

If "PQR" and "XYZ" are the congruent triangles, for example, the Definition tells us there is some correspondence under which the angles of PQR are equal to the corresponding angles of XYZ, and such that the pairs of sides subtending corresponding angles are equal as well (see Figure 30). The correspondence need not, however, be the alphabetical one that $\angle P$, $\angle Q$, $\angle R$ correspond in order to $\angle X$, $\angle Y$, $\angle Z$. Rather the correspondence might be (as the drawing suggests) that $\angle P = \angle Y$, $\angle Q = \angle X$, and $\angle R = \angle Z$, in

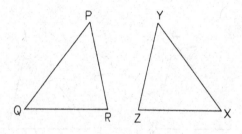

Figure 30

which case the equal pairs of subtending sides would be $QR = ZX$, $PR = YZ$, and $PQ = YX$.

Congruence of triangles consists of six equalities in all. The hypothesis of Theorem 4 gives us three of the six, and parts (1) and (3) of the conclusion give us the other three. Therefore, if we set conclusion (2) aside for the moment—we will return to it shortly—and streamline the hypothesis a bit, we get

Theorem 4 (SAS). *If two sides and the included angle of one triangle are equal respectively to two sides and the included angle of another triangle, then the triangles are congruent.*

This is the first Theorem to give us power to *infer* rather than power to *construct*. Here is Euclid's proof.

Proof.
1. Let "*ABC*" and "*DEF*" be the two triangles   hypothesis
   with, say, $AB = DE$, $AC = DF$, and $\angle BAC = \angle EDF$ (see Figure 31).
2. Pick up △*ABC* and set it down on △*DEF* so that
   point *A* coincides with point *D* and *AB* runs
   along *DE*.
3. *B* will coincide with *E*.   1 ($AB = DE$)
4. *AC* will run along *DF*.   1 ($\angle BAC = \angle EDF$)
5. *C* will coincide with *F*.   1 ($AC = DF$)
6. There can be only one straight line joining the   Post 1
   point $B = E$ to the point $C = F$.
7. Therefore *BC* will coincide with *EF*.   6
8. $\angle ABC = \angle DEF$, $\angle ACB = \angle DFE$, $BC = EF$ C.N. 4
9. Triangles *ABC* and *DEF* are congruent.   1, 8, Def "congruent"

Euclid's use of Postulate 1 in step 6 justifies our addition to that Postulate on p. 40.

In step 2 Euclid is using "superposition," a technique going back 300 years to Thales. In recent times, at least since 1550, superposition has been roundly criticized by mathematicians and philosophers alike as subversive of the nature of the *Elements*. The philosopher Arthur Schopenhauer, for example,

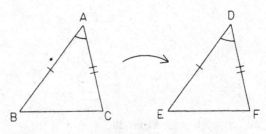

Figure 31

made this observation in 1818:

I am surprised that, instead of the fifth Postulate, the fourth Common Notion is not rather attacked: "Figures which coincide are equal to one another." For *coincidence* is either mere tautology, or something entirely empirical, which belongs, not to pure intuition, but to external sensuous experience.

—*The World as Will and Idea*, Book II

Others objected because they saw the essence of geometry to be the comparison of figures *at a distance*, a program that would be vitiated if figures were carried around and compared directly; and certainly the *applicability* of geometry to science and engineering derives from its conclusions about figures that are inaccessible.

The Greeks themselves may have had misgivings. Superposition seems to have been used less and less frequently as time went on. Euclid's own reluctance to use it in Book I is clear. There are theorems he *could* have proven by superposition, and such proofs would have been considerably shorter than the ones he actually presents, but he uses the technique only twice—here and in Theorem 8.

The proof has a formal problem as well. Apart from the question of whether we *want* to allow superposition, the fact is that no Common Notion or Postulate *does*. Euclid gives no reason for step 2. Commentators on the *Elements* are agreed that he probably interpreted Common Notion 4 as permitting superposition—this is why Schopenhauer linked the two in the passage quoted above—but even so step 2 is problematic because the Common Notion doesn't explicitly say that figures may be moved.[15]

My own opinion is that Euclid was in a bind. He *needed* the SAS congruence criterion, as it is fundamental to his geometric system. He could either make it an axiom and assume it, or make it a theorem and prove it. His readers would expect him to prove it, for in previous geometry books SAS had always been a theorem (proven by superposition). Besides, SAS *sounded* like a theorem—it didn't have the simplicity it seemed an axiom ought to have, and Euclid knew he was already in trouble on this score with his complicated Postulate 5. No, SAS would have to be a theorem. But he could devise no proof to replace the traditional superposition proof! He could have justified his use of superposition by postulating, unequivocally, that figures may be superposed, but he didn't want to endorse a technique that seemed on its way to obsolescence, and which in addition seemed inconsistent with the generally static character of his system. So he took what must have seemed the only way out: he used superposition, which he knew his readers would accept, but diplomatically called as little attention to it as possible.[16]

Complicated though the SAS criterion is, Hilbert and most other modern geometers make it an axiom. We will do the same.

Postulate 8 (SAS). If two sides and the included angle of one triangle are equal respectively to two sides and the included angle of another triangle, then the triangles are congruent.

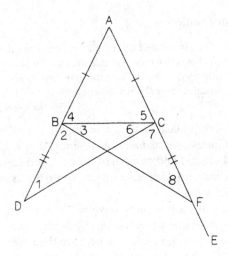

Figure 32

There's now a blank between Theorems 3 and 5. Theorem 4 has vanished, SAS has been promoted to postulate, and superposition has been banished from the system. Even so we should hang onto Euclid's superposition argument; it won't have the status of a proof, but it will be nice to have as an extra-systemic *explanation* we can show people who doubt the complicated new Postulate 8.

Theorem 5. *In isosceles triangles the angles at the base are equal to one another . . . .*

By Definition 20 an "isosceles" triangle has two equal sides; the "base" is just the third side. In his enunciation Euclid adds another clause about the angles "under the base" also being equal—in Figure 32, $\angle DBC = \angle FCB$—but I have omitted it and the steps that prove it.

It is amazing Euclid was able to prove this theorem so early! To see what I mean, close the book and try to invent a proof that uses only what we have done. Modern texts reserve this theorem until more tools are at hand.

Proof.

1. Let "ABC" be an isosceles triangle with, say,   hypothesis, Def
   $AB = AC$ (see Figure 32).   "isosceles"

(To show: $\angle ABC = \angle ACB$.)

2. Produce AB to "D."   Post 2
3. Produce AC to "E" so that $CE > BD$.   Post 2
4. From CE cut off C"F" $= BD$.   3, Th 3
5. Draw DC and FB.   Post 1
6. $AD = AF$ 1, 4, C.N. 2

7. $\angle A = \angle A$	C.N. 4
8. Triangles ABF and ADC are congruent.	1, 7, 6, SAS (Post 8)
9. $\angle 8 = \angle 1$, $BF = DC$	Def "congruent"
10. Triangles BCF and DBC are congruent.	4, 9, SAS
11. $\angle 4 + \angle 3 = \angle 5 + \angle 6$	8, Def "congruent"
12. $\angle 3 = \angle 6$	10, Def "congruent"
13. $\angle 4 = \angle 5$	11, 12, C.N. 3
14. In $\triangle ABC$ the angles at the base are equal to one another.	13

The unnumbered item I placed by itself, in parentheses, between steps 1 and 2 is the gist of a comment Euclid makes to indicate the direction the proof will take. It is of course not part of the proof and needs no reason. I will indicate future remarks, Euclid's or my own, in the same way. Also, when I number an angle, the number denotes the smallest angle containing the number in the finished drawing.

Theorem 5 is said to have been proven by Thales himself. His method, however, must have been different—there is persuasive evidence in Aristotle's *Prior Analytics* (I 24, 41 b 13–22) that the standard proof fifty years before Euclid had employed another, less rigorous, strategy, making it likely that the proof just rehearsed was fairly new when the *Elements* was composed, and may be Euclid's own.

In the late Middle Ages Theorem 5, accompanied by Euclid's figure and proof, came to be called the *Pons Asinorum*, "Bridge of Asses." The origin of this nickname seems to have had something to do with the theorem's perceived difficulty. The figure suggests a bridge and, depending on one's interpretation, the "asses" were either those who were sure-footed enough to make their way safely across, or those unable to proceed (see Figure 33).

Many people are intimidated by Euclid's proofs, and find proof-construction in even simple exercises bewildering work. In particular,

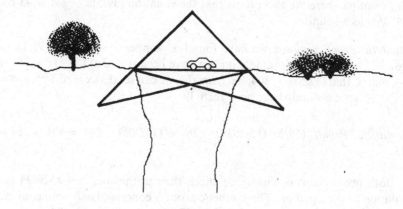

Figure 33

defeated-looking students frequently come to me and say, "I feel like one of those asses unable to proceed. I could *never* have thought up that proof for Theorem 5." I tell them I'm not sure I could have either, given the maze of auxiliary points and lines, most of them useless, that *could* have been constructed after step 1.

The cause of all this difficulty is Euclid's manner of presenting demonstrations. Euclid's geometry is called "synthetic" geometry, as opposed to the "analytic" geometry published by René Descartes[17] in 1637 and taught today in high school algebra courses. In analytic geometry straight lines and circles are represented by algebraic equations, a fact that misleads many students into thinking that the term "analytic" is a synonym for "algebraic." While it is true that Descartes' geometry is "analytic" as a consequence of its algebraic method, the term "analytic" as used in mathematics is much more general and antedates Descartes' symbolic algebra by almost two millenia. It refers to any method of *working backward* by *separating* a statement whose proof is desired into components that are logically *prior*. "Synthesis," the companion term, means proof by *combining* separate elements into a deductive sequence that *culminates* in the statement to be proved.

Pappus—a Greek geometer who worked in Alexandria 600 years after Euclid, and compiled an immense anthology of advanced geometry— explained the distinction this way (quoted in Heath's *Euclid*, p. 138):

> ... in analysis we assume that which is sought as if it were [already] done, and we inquire what it is from which this results, and again what is the antecedent cause of the latter, and so on, until by so retracing our steps we come upon something already known or belonging to the class of first principles, and such a method we call analysis as being solution backwards.
>
> But in *synthesis*, reversing the process, we take as already done that which was last arrived at in the analysis and, by arranging in their natural order as consequences what were before antecedents, and successively connecting them with one another, we arrive finally at the construction of what was sought; and this we call synthesis.

For example, here are two proofs that the equation $(39/5)x - 261 = 43$ has $1520/39$ as a solution.

Analytic Proof. Suppose we have found a number x such that $(39/5)x - 261 = 43$. Then for this same x it must have been true that $(39/5)x = 261 + 43$, that is, that $(39/5)x = 304$; but from this we can see that $x = (5/39) \cdot 304 = 1520/39$, so x can only have been $1520/39$.

Synthetic Proof. $(39/5) \cdot (1520/39) - 261 = (1520/5) - 261 = 304 - 261 = 43$.

Both proofs convince us, if we check their arithmetic, that $1520/39$ is a solution to the equation. The synthetic proof is concerned *only* with convincing us of that fact, so it proceeds as expeditiously as possible. The analytic proof is longer because it also explains *how* the fact was discovered. Which proof I use therefore depends on my motive.

High school algebra texts usually recommend that a student construct both proofs, the synthesis as a check that the analysis was correctly done. However most students quickly discover that in algebra analysis is practically foolproof and abandon the synthetic check as redundant. Today's scientific and engineering mathematics, being mostly algebraic in form, is dominated by analysis.

In Greek mathematics, however, synthesis was dominant. Greek mathematics was geometric in form, and analysis is less reliable in that context. It does not proceed as automatically, with every implication reversible, as it does in solving an algebraic equation. Doubtless the Greeks used some sort of (presumably) nonalgebraic analysis to *discover* their proofs—no mathematician can work forward all the time—but they did not consider a proof finished until it had been completely recast in synthetic form.

Thus the formidability of Euclid's proofs. They are syntheses, with the analyses *omitted*. The scaffolding has been removed and they stand, like the pyramids, leaving us to wonder how they were constructed.

The next theorem is the first in the *Elements* to be proven by contradiction. It is the "converse" of Theorem 5.

The *converse* of a conditional statement is obtained by interchanging the hypothesis and conclusion. The converse of "If H, then C" is "If C, then H." Though they can seem to say almost the same thing, a statement and its converse say things which, logically, are very different. For example, the converse of "If one is a woman then one is a human being" is "If one is a human being then one is a woman." For this reason the converse of a theorem need not be true and, even if it is, requires its own proof. Theorem 6 is the converse of Theorem 5 because in Theorem 6 were are given a triangle with two equal angles (Theorem 5's conclusion) and conclude that it is isosceles (Theorem 5's hypothesis).

Theorem 6. *If in a triangle two angles be equal to one another, the sides which subtend the angles will also be equal to one another.*

Proof.

1. Let "*ABC*" be a triangle with, say, $\angle ACB =$ hypothesis
   $\angle ABC$ (see Figure 34).
(To show: $AB = AC$.)

2. Pretend $AB \neq AC$, say $AB > AC$.   RAA hypothesis
3. From *BA* cut off $B"D" = CA$.   Th 3
4. Draw *DC*.   Post 1
5. $BC = BC$ C.N. 4
6. Triangles *DBC* and *ACB* are congruent.   3, 1, 5, SAS
7. area of $\triangle DBC =$ area of $\triangle ACB$ 6

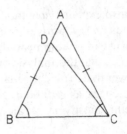

Figure 34

8. But area of △*ACB* > area of △*DBC*.   C.N. 5
9. Contradiction.   7 and 8
10. Therefore *AB = AC*.   2–9, logic

In step 2, in order to be specific, Euclid took $AB > AC$, so the contradiction in step 9 actually implies only that AB is not greater than AC. But he knew an argument similar to steps 2–9 could be inserted between steps 9 and 10, showing that AB is not less than AC either.

In part (2) of the conclusion to Theorem 4 (p. 53) Euclid clearly stated that congruent triangles have equal areas, but in our revision (p. 54) we dropped that clause to bring the theorem (now Postulate 8) into line with modern textbooks. As a result, while Euclid made no tacit assumption in step 7—it follows perfectly well from *his* SAS–*we* did. (The tables have turned!) Let us hasten, then, to reinstate the stricken clause.

Postulate 9. Congruent triangles have equal areas.

Now we can change the reason for step 7 to "6, Post 9."

Theorem 7. *Given two straight lines constructed on a straight line [from its extremities] and meeting in a point, there cannot be constructed on the same straight line [from its extremities], and on the same side of it, two other straight lines meeting in another point and equal to the former two respectively, namely each to that which has the same extremity with it.*

This will be used in the proof of Theorem 8. What it says, succinctly, is that the situation depicted in Figure 35 is impossible. Contrast that with Figure 36, which represents two situations that not only can occur, but frequently do.

Euclid. ... Theorem 7 shows that, of all plane figures that can be made by hingeing rods together, the *three*-sided ones (and these only) are *rigid* (which is another way of stating the fact that there cannot be *two* such figures on the same base).

. . . .

Minos. You have made out a good case. ... It is one of the very few Theorems that have a direct bearing on practical science. I have often found pupils much interested in

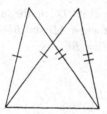

Figure 35

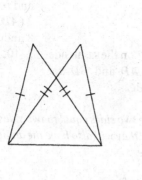

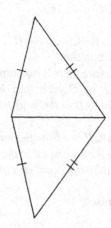

Figure 36

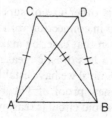

Figure 37

learning that the principle of the rigidity of triangles is of constant use in architecture, and even in so homely a matter as the making of a gate.

—*Euclid and His Modern Rivals*, Act IV, Scene 6

Proof.

1. Let "*AC*" and "*B*"*C* be the two straight lines   hypothesis
   constructed on the straight line *AB* from its
   extremities and meeting in the point *C* (see
   Figure 37).

2. Pretend there is another point "D" on the same side of AB as C such that when AD and BD are drawn, $AD = AC$ and $BD = BC$.   RAA hypothesis

3. Draw DC.   Post 1
4. $\angle ADC < \angle BDC$ C.N. 5
5. $\angle BDC = \angle BCD$ 2 ($BD = BC$), Th 5
6. $\angle ADC < \angle BCD$ 4, 5, C.N. 6 (If $a < b$ and $b = c$ than $a < c$.)

7. $\angle BCD < \angle ACD$ C.N. 5
8. $\angle ADC < \angle ACD$ 6, 7, C.N. 6 (If $a < b$ and $b < c$ than $a < c$.)

9. But $\angle ADC = \angle ACD$.   2 ($AD = AC$), Th 5
10. Contradiction.   8 and 9
11. Therefore there is no point "D" on the same side of AB as C such that when AD and BD are drawn, $AD = AC$ and $BD = BC$.   2–10, logic

Theorem 8. *If two triangles have the two sides equal to two sides respectively, and have also the base equal to the base, they will also have the angles equal which are contained by the equal straight lines.*

Rephrased:

Theorem 8 (SSS). *If the three sides of one triangle are equal respectively to the. three sides of another triangle, then the triangles are congruent.*

Euclid's proof involves his other use of superposition, and presents the same difficulties we discussed in connection with his proof of Theorem 4. We could avoid them, as we did in that case, by making the Theorem an axiom. This time, however, there is a less drastic measure that will suffice. We will postpone proving Theorem 8 until we have established Theorem 23, then use Theorem 23 to construct a new proof of Theorem 8 that does not involve superposition. Of course, to avoid circular reasoning, this means we will have to refrain from using SSS in the meantime.

Presently we have the power to do five things: draw straight lines (Postulate 1), prolong them (Postulate 2), draw circles (Postulate 3), construct equilateral triangles (Theorem 1), and copy finite straight lines (Theorems 2 and 3). In the next four theorems and an accompanying postulate (our last) we acquire four new powers.

Theorem 9. [*It is possible*] *to bisect a given rectilinear angle.*

We who have been trained to think of angles as measured by numbers, which exist in their own realm apart from geometry, find it particularly

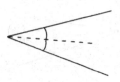

Figure 38

obvious that an angle can be cut into two, three, or however many equal pieces because the corresponding numbers can be immediately divided by 2, 3, or any other number. To Euclid, however, the measure of an angle—he would have said its "magnitude"—was not representable outside geometry. To "bisect" an angle was to cut it in half *geometrically* by a straight line through the vertex (see Figure 38). Since geometry's existence criterion was constructability with compass and straightedge, and it is not obvious those tools are adequate to construct the straight line in question, this theorem was, in Euclid's time, as worthy of careful proof as any other. (In fact, when the *Elements* was composed Greek mathematicians had been trying without success to *tri*sect an angle—cut it in thirds—for years! In the 19th century it was finally shown that with compass and straightedge only, most angles *cannot* be trisected, and Euclid's existence criterion came to be seen as too restrictive. Under modern existence criteria angles can be cut into any number of equal pieces.)

Euclid begins his proof by choosing a point "at random" on one side of the angle. He does a similar thing in the proof of Theorem 12. Metaphorically he is setting down one leg of his compass, but not quite at random because there are restrictions—here it has to land on a given straight line, and in Theorem 12 *off* a given infinite straight line but on a given side of it. We will acknowledge this new power of the compass by framing a new postulate, our last. (Of course Euclid has been taking this power for granted all along. What's new is our realization that he has.)

Postulate 10. It is possible to choose a point at random

 (i) in the plane,
 (ii) between the endpoints of a given finite line,
 (iii) outside or inside a given circle (or triangle), or
 (iv) on a given side of a finite or infinite straight line.

Proof of Theorem 9.
  1. Let "*BAC*" be the given rectilinear angle (see   hypothesis
     Figure 39).
  2. Choose a point "*D*" at random on *AB*.   Post 10 (ii)
  3. Produce *AC*, if necessary, to make it longer than   Post 2
     AD.
  4. From *AC*, or *AC* produced, cut off *A*"*E*" = *AD*.   Th 3

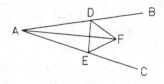

Figure 39

5. Draw *DE*.   Post 1
6. On *DE* construct an equilateral triangle *DE*"*F*".   Th 1
7. Draw *AF*.   Post 1
8. *DF* = *EF*   Def "equilateral
                                                              triangle"

(At this point Euclid says triangles *ADF* and *AEF* are congruent by SSS. As we are postponing the proof of SSS we cannot follow him, so will use SAS instead.)

9. ∠*ADE* = ∠*AED*   4, Th 5
10. ∠*FDE* = ∠*FED*   8, Th 5
11. ∠*ADF* = ∠*AEF*   9, 10, C.N. 2
12. Triangles *ADF* and *AEF* are congruent.   4, 11, 8, SAS
13. ∠*DAF* = ∠*EAF*   Def "congruent"
14. *AF* bisects ∠*BAC*.   13

Theorem 10. [*It is possible*] *to bisect a given finite straight line.* (See Figure 40.)

Proof.
1. Let "*AB*" be the given finite straight line (see   hypothesis
   Figure 41).

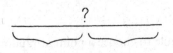

Figure 40

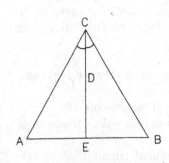

Figure 41

2. On AB construct an equilateral triangle $AB"C"$.   Th 1
3. Draw $C"D"$ to bisect $\angle ACB$.   Th 9
4. Produce CD until it intersects AB in a point $"E"$.   Post 2, Post 6 (iii), Post 1

5. $AC = BC$ Def "equilateral triangle"

6. $CE = CE$ C.N. 4
7. Triangles ACE and BCE are congruent.   5, 3 ($\angle ACE = \angle BCE$), 6, SAS

8. $AE = EB$ Def "congruent"
9. E bisects AB.   8

Note that while the "bisector" of an angle is a straight line, the "bisector" of a finite straight line is just a point.

You may have been puzzled by the reason for step 4. When late 19th-century geometers went over the *Elements* with a fine-toothed comb looking for logical crevices, one dismaying discovery was of how frequently Euclid was guilty of reasoning from the diagram. The very presence of geometrical diagrams is in fact the reason so many of Euclid's tacit assumptions lay unnoticed for so long. Using a diagram to follow an argument, we (Euclid too) unconsciously accept as proven any extra assertions it may make about the things we are studying. These assertions are subtle because they are made, and accepted, without ever being put into words. Step 4 is a good example. With Figure 41 before us it is obvious that if CD is produced it will intersect AB. But let's back up a bit. After CD bisects $\angle ACB$ in step 3 we have the situation depicted in Figure 42. With only Euclid's original axioms, how do we know CD produced won't (for example) spiral around inside the triangle, without ever intersecting it a second time? (See Figure 43.) Our intuitions thunder, "Because CD is straight!" But "straight line" is a primitive term, so we don't officially know what "straight" means. Any extra properties we want straight lines to have will have to be spelled out in new axioms. In this case the spelling-out was done in 1882 when Postulate 6 (iii) was uncovered by Moritz Pasch.[18] But though Postulate 6 (iii) now guarantees that CD, if produced sufficiently, will intersect the triangle a second time, it doesn't say *where* (see Figure 44).

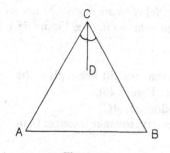

Figure 42

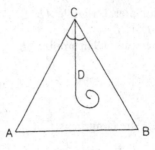

Figure 43

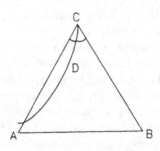

Figure 44

Why between A and B, instead of between A and C, say, or at A? Because we know from our revised Postulate 1 that only *one* straight line can join two given points. If CD produced met AC between A and C or at A, there would be two straight lines joining C to the point of intersection; and similarly CD produced cannot meet BC either.

Our earlier discussions involving the notion of "right angle" were contingent, for their significance, on the *existence* of right angles, something that so far has not been explicitly established, though a pair did make an unacknowledged appearance in the proof of Theorem 10. Theorem 11 exploits the same idea to tie up this loose end.

Theorem 11. [*It is possible*] *to draw a straight line at right angles to a given straight line from a given point on it.* (See Figure 45.)

Proof.
1. Let "AB" be the given straight line with C the   hypothesis
   given point on it (see Figure 46).
2. Choose "D" at random on AC.   Post 10 (ii)
3. Produce CB, if necessary, to make it longer than   Post 2
   DC.

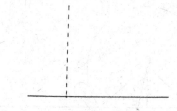

Figure 45

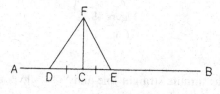

Figure 46

4. From *CB*, or *CB* produced, cut off *C"E" = DC*.   Th 3
5. On *DE* construct an equilateral triangle *"F" DE*.   Th 1
6. Draw *FC*.   Post 1
7. *FD = FE*   Def "equilateral
triangle"

(Here Euclid uses SSS to say triangles *FDC* and *FEC* are congruent. We will avoid this, as before, with SAS.)

8. ∠ *FDC* = ∠ *FEC*   Th 5
9. Triangles *FDC* and *FEC* are congruent.   7, 8, 4, SAS
10. ∠ *FCD* = ∠ *FCE*   Def "congruent"
11. *FC* is at right angles to *AB*.   10, Def "right angle"

Theorem 12. *To a given infinite straight line, from a given point which is not on it,* [*it is possible*] *to draw a perpendicular straight line.*

Having just learned to construct a perpendicular "from the bottom up," we now learn to do so "from the top down." (See Figure 47.) Euclid makes the straight line infinite to be sure the portion on which the perpendicular will land has already been drawn.

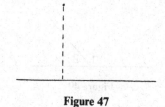

Figure 47

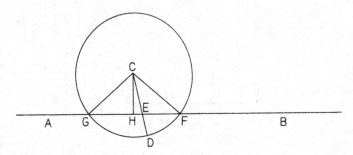

Figure 48

Proof.
1. Let "*AB*" be the infinite straight line and "*C*"   hypothesis
   the point not on it (see Figure 48).
2. Choose a point "*D*" at random on the other side   Post 10 (iv)
   of *AB*.
3. Draw *CD*.   Post 1
4. *CD* intersects *AB* in a point "*E*".   Post 7 (ii)
5. Draw a circle with center *C* and radius *CD*.   Post 3
6. *EB* intersects the circle in a point "*F*".   Post 6 (ii)
7. *FE* produced (i.e., *FA*) intersects the circle in a Post 6 (iii)
   point "*G*".
8. Bisect *GF* at "*H*".   Th 10
9. Draw *CG*, *CF*, and *CH*.   Post 1
10. *CG* = *CF*   Def "radius"
(Once more Euclid uses SSS, but once more we can use SAS instead.)
11. ∠*CGH* = ∠*CFH*   Th 5
12. Triangles *CGH* and *CFH* are congruent.   10, 11, 8, SAS
13. ∠*CHG* = ∠*CHF*   Def "congruent"
14. *CH* is perpendicular to *AB*.   Def "perpendicular"

Theorem 13. *If a straight line set up on a straight line make angles, it will make either two right angles or angles equal to two right angles.*

That is, in Figure 49, angles 1 and 2 are either right angles themselves or, if not, have the same sum that two actual right angles would have. Euclid's

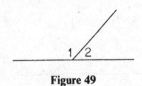

Figure 49

careful distinction between the possibilities reflects his attitude that an angle's magnitude is somehow attached to the specific angle itself and cannot be discussed independently as we would a number. The distinction, no longer considered important, is left behind if we rephrase the theorem in the customary way: "Supplementary[19] angles add up to 180°."

Though the *significance* of Theorem 13 depends on Theorem 11 and Postulate 4—because it requires the existence and constancy of 180° as a standard to which angle-sums can be referred—you may think we have had possession of at least the bald *truth* of the theorem since the Definition of "right angle." Admittedly, if the straight line set up on the other makes angles that are equal, we have; but should the angles be unequal, we have not. Only with Theorem 11 have we acquired the tool crucial to inferring that the space occupied by two unequal angles can also be occupied by two right angles.

Proof.

1. Let "AB" be the straight line set up on the   hypothesis
   straight line "CD", making the angles ABC and
   ABD (see Figure 50).

2. If $\angle ABC = \angle ABD$ then AB makes two right   Def "right angle"
   angles.

3. If $\angle ABC \neq \angle ABD$, say $\angle ABC > \angle ABD$,   Th 11
   draw B"E" at right angles to CD.

4. $\angle ABC = \angle 1 + \angle 2, \angle 3 = \angle 3$ C.N. 4

5. $\angle ABC + \angle 3 = \angle 1 + \angle 2 + \angle 3$ 4, C.N. 2

6. $\angle 2 + \angle 3 = \angle EBD, \angle 1 = \angle 1$ C.N. 4

7. $\angle 1 + \angle 2 + \angle 3 = \angle 1 + \angle EBD$ 6, C.N. 2

8. $\angle ABC + \angle 3 = \angle 1 + \angle EBD$ 5, 7, C.N. 1

9. $\angle 1$ and $\angle EBD$ are right angles.   3

10. Therefore, AB makes angles equal to two right   8, 9
    angles.

Neither Theorem 14, which is the converse of Theorem 13, nor Theorem 15, attributed to Thales, is difficult to prove. For each I'll start the proof and leave the rest as an exercise.

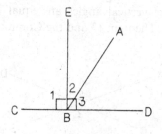

Figure 50

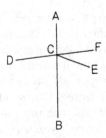

Figure 51

Theorem 14. *If with any straight line, and at a point on it, two straight lines not lying on the same side make the adjacent angles equal to two right angles, then the two straight lines will be in a straight line with one another.*

Proof.
1. Let "*AB*" be the straight line and "*C*" the point   hypothesis
   on it at which the two straight lines "*D*"*C* and
   "*E*"*C*, not lying on the same side of *AB*, make the
   adjacent angles *DCA* and *ECA* add up to 180°.
   (See Figure 51.)
(To show: *DCE* is actually a single straight line. Euclid's proof is by contradiction.)
2. Pretend *DCE* is not a single straight line.   RAA hypothesis
3. Produce *DC* in a straight line to "*F*".   Post 2
(Now use Theorem 13 and some Common Notion-based computations to derive a contradiction.)

Theorem 15. *If two straight lines cut one another, they make the vertical angles equal to one another.*

Proof.
1. Let "*AB*" and "*CD*" be the straight lines that   hypothesis
   cut one another at "*E*" (see Figure 52).
(When Euclid says "the vertical angles are equal" he means that ∠1 = ∠2 and ∠3 = ∠4. Use Theorem 13 and the Common Notions.)

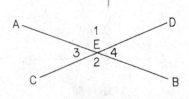

Figure 52

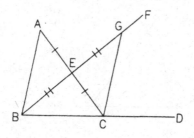

Figure 53

Theorem 16. *In any triangle, if one of the sides be produced, the exterior angle is greater than either of the interior and opposite angles.*

Proof.
1. Let "*ABC*" be the triangle with, say, *BC* pro-   hypothesis
   duced to "*D*" (see Figure 53).

(The "exterior angle" is then ∠*ACD* and the "interior and opposite angles" ∠*ABC* and ∠*BAC*. First we will show ∠*ACD* > ∠*BAC*.)

2. Bisect *AC* at "*E*".	Th 10
3. Draw *BE*.	Post 1
4. Produce *BE* to "*F*" so that *EF* > *BE*.	Post 2
5. From *EF* cut off *E"G"* = *BE*.	Th 3
6. Draw *GC*.	Post 1
7. ∠*AEB* = ∠*GEC*	Th 15
8. Triangles *AEB* and *GEC* are congruent.	2, 7, 5, SAS
9 ∠*BAC* = ∠*ACG*	Def "congruent"
10. ∠*ACD* > ∠*ACG*	C.N. 5
11. ∠*ACD* > ∠*BAC*	9, 10, C.N. 6 (If *a* = *b* and *c* > *b* then *c* > *a*.)

I leave it to you to complete the proof by showing ∠*ACD* > ∠*ABC*. Start by using Post 2 to produce *AC* to "*H*" (Figure 54), imitate steps 2–11 to show ∠*BCH* > ∠*ABC*, then use this to obtain the conclusion.

Theorem 16 is called the "Exterior Angle Theorem." It is frequently useful, though the opportunities are easy to overlook.

Theorem 17. *In any triangle two angles taken together in any manner are less than two right angles.*

That is, the sum of any two angles of a triangle is less than 180°. In the context of a straight line falling on two others (Figure 55) this is the converse of Postulate 5. Postulate 5 says that if ∠1 + ∠2 < 180° then the straight lines will meet, forming a triangle. Theorem 17 says that if a triangle is formed then ∠1 + ∠2 < 180°.

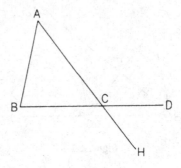

Figure 54

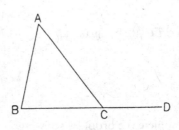

Figure 55

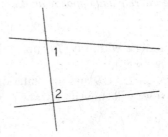

Figure 56

Proof.
1. Let "*ABC*" be a triangle with, say ∠*ABC* and   hypothesis
    ∠*ACB* the two angles (see Figure 56).
2. Produce *BC* to "*D*".   Post 2
(The rest of the proof is an exercise.)

Theorems 18 and 19 complete the stories begun in Theorems 5 and 6 respectively. In Theorem 5 we were given that two sides of a triangle were equal, and in Theorem 6 that two angles were equal; Theorems 18 and 19 tell us what can be deduced if two sides or two angles are *unequal*.

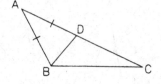

Figure 57

Theorem 18. *In any triangle the greater side subtends the greater angle.*

Proof.
1. Let "*ABC*" be a triangle with, say, *AC > AB*   hypothesis
   (see Figure 57).
(To show: ∠*ABC* > ∠*ACB*.)
2. From *AC* cut off *A"D" = AB*.   Th 3
3. Draw *BD*.   Post 1
4. ∠*ABC* > ∠*ABD*   C.N. 5
5. ∠*ABD* = ∠*ADB*   Th 5
6. ∠*ABC* > ∠*ADB*   4, 5, C.N. 6 (If *a > b*
                                                        and *b = c* then *a > c*.)
7. ∠*ADB* > ∠*ACB*   Th 16 (△*DBC*)
(Use of Theorem 16 can be hard to spot, so I mention the triangle. The exterior
angle *ADB* is greater than the interior and opposite angle *ACB*.)
8. Therefore ∠*ABC* > ∠*ACB*.   6, 7, C.N. 6 (If *a > b*
                                                        and *b > c* than *a > c*.)

Theorem 19. *In any triangle the greater angle is subtended by the greater side.*

This is of course the converse of Theorem 18. The two are easily confused
because this theorem's passive construction ("is subtended by") tempts us to
interpret it as Theorem 18. Just remember that in each theorem the first thing
mentioned is the hypothesis.

Proof.
1. Let "*ABC*" be a triangle with, say ∠*ABC* >   hypothesis
   ∠*ACB* (see Figure 58).
(To show: *AC > AB*.)
2. Pretend *AC = AB*.   RAA hypothesis
3. Then ∠*ABC* = ∠*ACB*.   Th 5
4. Contradiction.   1 and 3
5. Therefore *AC ≠ AB*.   2–4, logic
6. Now pretend *AC < AB*.   RAA hypothesis
7. Then ∠*ABC* < ∠*ACB*.   Th 18
8. Contradiction.   1 and 7

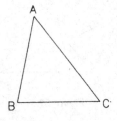

Figure 58

9. Therefore $AC \not< AB$.   6–8, logic
10. Therefore $AC > AB$.   5, 9, C.N. 6 (Either
                                                     $a = b; a < b$, or
                                                     $a > b$.)

This kind of proof is called proof "by double contradiction." Euclid wants $AC > AB$. Unable to prove this directly, he sets out to disprove the alternatives instead. But pursuing them leads him down dissimilar paths, so both contradictions have to be explicitly derived.

Theorem 20. *In any triangle two sides taken together in any manner are greater than the remaining one.*

That is, the sum of any two sides of a triangle is greater than the third side. This theorem is called the "Triangle Inequality."

Proof.
1. Let "*ABC*" be a triangle with, say, *AB* and *AC*   hypothesis
   the two sides (see Figure 59).
(We have to show $AB + AC > BC$.)
2. Produce *BA* to "*D*" so that $AD > AC$.   Post 2

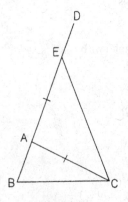

Figure 59

3. From AD cut off $A``E" = AC$.   Th 3
4. Draw EC.   Post 1
(The rest of the proof is an exercise.)

Theorem 20 is the closest Euclid ever comes to saying "a straight line is the shortest path between two points," though this is often ascribed to him (p. 31).

It was the habit of the Epicureans, says Proclus, to ridicule this theorem as being evident even to an ass and requiring no proof, and their allegation that the theorem was "known" even to an ass was based on the fact that, if fodder is placed at one angular point and the ass at another, he does not, in order to get his food, traverse the two sides of the triangle but only the one side separating them .... Proclus replies truly that a mere perception of the truth of the theorem is a different thing from a scientific proof of it and a knowledge of the reason *why* it is true. Moreover ... the number of axioms should not be increased without necessity.

—Sir Thomas L. Heath in *Euclid*, p. 287

(Proclus, A.D. 410–485, was a neo-Platonic philosopher who wrote a commentary on Book I of the *Elements*.)

I'm going to pass over Theorems 21 and 22 because we won't be needing them. This brings us to

Theorem 23. *On a given straight line and at a point on it [it is possible] to construct a rectilinear angle equal to a given rectilinear angle.*

This is the angle analog of Theorem 3: it gives us the power to make a copy, here, of an angle located elsewhere (see Figure 60). Recall (p. 33) that "rectilinear" means only than an angle is made up of two *straight* lines (as always), contributing nothing to our discussion.

Theorem 23 is the theorem we need to fashion a superposition-free proof of SSS (Theorem 8). Ironically, Euclid's proof of Theorem 23 depends on SSS, and this time the dependence is basic! Before—in Theorems 9, 11, and 12—we were able to circumvent Euclid's use of SSS by only slightly modifying his proofs; now we will have to substitute an entirely new proof. It is long, but never difficult.

Proof.
1. Let "AB" be the given straight line, "C" the   hypothesis
   point on it, and "DEF" the given angle (see
   Figure 60).

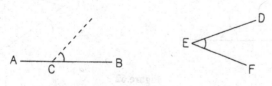

Figure 60

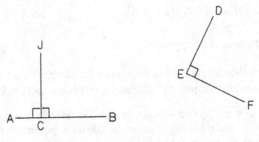

Figure 61

(What we have to show is that we can construct a straight line through *C* making an angle equal to ∠ *DEF*. The proof is really three separate proofs— one for the case that ∠ *DEF* is right, one for the case that it is acute, and one for the case that it is obtuse.)

Case 1. ∠ DEF is right (Figure 61).

  2. Draw *C"J"* at right angles to *AB*.   Th 11

  3. ∠ *JCB* = ∠ *DEF*   hypothesis of Case 1,
                                           2, Post 4

(When a proof is broken up into cases, each case provides an extra hypothesis that is used only within that case. Thus what we have actually proved thus far is this: If ∠ *DEF* is right, we can construct an equal angle *JCB*.)

Case 2. ∠ DEF is acute (Figure 60).

  4. Draw *D"G"* perpendicular to *EF* (produced if   Th 12 (Post 2)
     necessary).

(It may be necessary to produce *EF* to include the point *G* on which the perpendicular will land. Intuitively we recognize that *G* will be on the same side of *E* as *F*, as drawn in Figure 64, but for completeness steps 5–20 verify that this is the case.)

  5. Pretend *G* is on the other side of *E* from *F*   RAA hypothesis
     (see Figure 62).

  6. ∠ *DGE* = 90°   4

  7. ∠ *DEF* > ∠ *DGE*   Th 16

  8. ∠ *DEF* > 90°   6, 7, C.N. 6 (If *a* = *b*
                                           and *c* > *a* then *c* > *b*.)

Figure 62

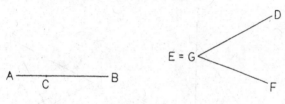

Figure 63

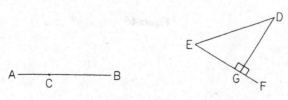

Figure 64

9. But $\angle DEF < 90°$.	hypothesis of Case 2
10. Contradiction.	8 and 9
11. Therefore G is not on the other side of E from F.	5–10, logic
12. Now pretend G coincides with E (see Figure 63).	RAA hypothesis
13. DG coincides with DE.	Post 1
14. $\angle DGF = \angle DEF$	C.N. 4
15. $\angle DGF = 90°$	4
16. $\angle DEF < 90°$	hypothesis of Case 2
17. $\angle DGF > \angle DEF$	15, 16, C.N. 6 (If $a = b$ and $c < b$ then $a > c$.)
18. Contradiction.	14 and 17
19. Therefore G does not coincide with E.	12–18, logic
20. Therefore G is on the same side of E as F (see Figure 64).	11, 19

(Now we can get on with our proper job, constructing a copy of $\angle DEF$.)

21. From CB, or CB produced, cut off $C"H" = EG$ (see Figure 65).	Th 3 (Post 2)
22. Through H draw a straight line at right angles to AB, and produce it to "I" so that $HI > DG$.	Th 11, Post 2
23. From HI cut off $H"J" = DG$.	Th 3
24. Draw CJ.	Post 1

($\angle JCB$ is the desired copy of $\angle DEF$, as we will now show.)

25. $\angle JHC = \angle DGE$	4, 22, Post 4
26. Triangles JCH and DEG are congruent.	21, 25, 23, SAS
27. $\angle JCB = \angle DEF$	Def "congruent"

(What Case 2 proves is this: If $\angle DEF$ is acute, we can construct an equal angle JCB.)

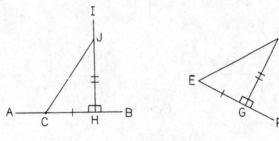

Figure 65

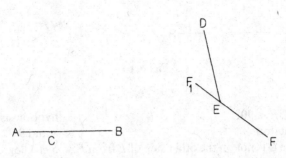

Figure 66

Case 3. ∠ *DEF is obtuse* (see Figure 66).

28. Produce *FE* to a point "F_1".	Post 2
29. ∠ *DEF*$_1$ + ∠ *DEF* = 180°	Th 13
30. ∠ *DEF* > 90°	hypothesis of Case 3
31. ∠ *DEF*$_1$ < 90°	29, 30, C.N. 6 (If $a + b = c$ and $b > \frac{1}{2}c$ then $a < \frac{1}{2}c$.)
32. Draw *C"J"* so that ∠ *JCA* = ∠ *DEF*$_1$ (see Figure 67).	31, Case 2

(Case 2 is a mini-theorem in its own right, and has already been proved. It says

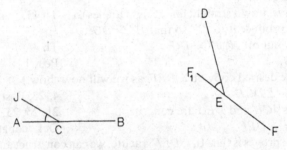

Figure 67

that if the angle we want to copy is acute—think of $\angle DEF_1$ as the angle we want to copy—then we can draw a straight line CJ that does the trick. Furthermore, Case 2 has a flexibility that enables us to make CA, instead of CB, the second side of the copy. In our proof of Case 2 we arranged things so that CB would be the second side—the decision was made in step 21—but we could just as well have given the nod to CA.)

33. $\angle JCA + \angle JCB = 180°$ Th 13

34. $\angle JCA + \angle JCB = \angle DEF_1 + \angle DEF$ 33, 29, C.N. 1

35. $\angle JCB = \angle DEF$ 34, 32, C.N. 3

(Case 1 proved: If $\angle DEF$ is right, we can construct an equal angle JCB. Case 2 proved: If $\angle DEF$ is acute, we can construct an equal angle JCB. Now Case 3 proves: If $\angle DEF$ is obtuse, we can construct an equal angle JCB. As $\angle DEF$ can only be right, acute, or obtuse, we have considered every possibility. Therefore, in *any* case, we can construct an equal angle JCB, and we have proven Theorem 23.)

At last we can prove SSS.

Theorem 8 (SSS). *If the three sides of one triangle are equal respectively to the three sides of another triangle, then the triangles are congruent.*

Proof.

1. Let "ABC" and "DEF" be the two triangles   hypothesis
   with, say, $AB = DE$, $BC = EF$, and $AC = DF$
   (see Figure 68).

(The idea is to get the triangles congruent by SAS. In view of step 1 all we need is the equality of one pair of corresponding angles.)

2. Pretend $\angle ABC \neq \angle DEF$,   say $\angle ABC >$ RAA hypothesis
   $\angle DEF$

3. On the straight line BC and at the point B, draw   Th 23
   $B"G"$ so that $\angle GBC = \angle DEF$.

4. $\angle ABC > \angle GBC$ 2, 3, C.N. 6 (If $a > b$
                                         and $c = b$ then $a > c$.)

5. BG is between BA and BC.   4

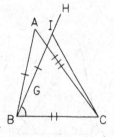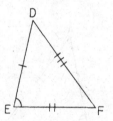

Figure 68

6. Produce *BG* to "*H*" so that *BH* > *DE*.	Post 2
7. From *BH* cut off *B*"*I*" = *DE*.	Th 3
8. Draw *CI*.	Post 1
9. Triangles *IBC* and *DEF* are congruent.	7, 3, 1 (*BC* = *EF*), SAS
10. *A* and *I* are different points.	5, Post 1
11. *BA* = *BI*	1 (*BA* = *DE*), 7, C.N. 1
12. *CI* = *DF*	9, Def "congruent"
13. *CA* = *CI*	1 (*CA* = *DF*), 12, C.N. 1

(Before reading the rest of the proof you may want to reread the statement of Theorem 7 and its explanation, pp. 60–61.)

14. Given two straight lines (*BA* and *CA*) constructed on a straight line (*BC*) from its extremities and meeting in a point (*A*), we have constructed on the same straight line (*BC*), from its extremities and on the same side of it, two other straight lines (*BI* and *CI*) meeting in another point (*I*) and equal to the former two respectively, namely each to that which has the same extremity with it (*BA* = *BI* and *CA* = *CI*) (see Figure 68).	7, 8, 10, 11, 13
15. Contradiction.	14 and Th 7
16. Therefore ∠*ABC* = ∠*DEF* (see Figure 69).	2–15, logic
17. Therefore triangles *ABC* and *DEF* are congruent.	1 (*AB* = *DE*, *BC* = *EF*), 16, SAS

This is probably a good time to talk about forcing points or lines to do multiple duty, an error people sometimes fall into. In a situation like Figure 70, for example, in which it is given that *AB* = *DC* and *AD* = *BC*, and required to prove that ∠*A* = ∠*C*, I have seen people (who must have forgotten SSS and remembered only SAS) proceed as follows:

"Draw straight line *BD* to bisect ∠*ABC* and   Th 9"
∠*ADC*.

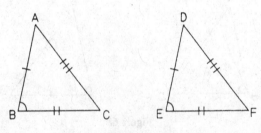

Figure 69

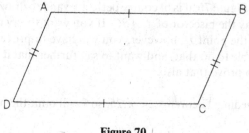

Figure 70

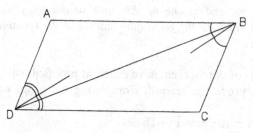

Figure 71

While it is certainly true that we can draw straight line *BD* (using Postulate 1) and that we can bisect ∠*ABC* and ∠*ADC* (using Theorem 9 twice), it is extremely unlikely that even two of these straight lines will coincide, let alone all three. In general the result of constructing all three would look something like Figure 71. The person has taken three straight lines—which admittedly exist—and *identified* them, forcing straight line *BD* to do triple duty.

Here's an example of the kind of logical hot water this sort of thing can lead to. Consider ∠*ABC* in Figure 72. Through the point *A* I can draw a perpendicular straight line, using Theorem 11, and through the point *C* I can draw a perpendicular straight line, using Theorem 11 again. I will identify these lines by saying

"Draw straight line *AC* perpendicular to *AB* and   Th 11
BC."

But now the resulting triangle has two right angles, in conflict with Theorem 17.

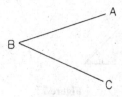

Figure 72

Returning to Figure 70, it is of course perfectly valid to draw *one* of the three straight lines, say the bisector of $\angle ABC$. If you want to say this straight line passes through the point D, however, you will have to prove that it does so; and if you are able to do that, and want to say further that it bisects $\angle ADC$, you will have to prove that also.

We won't be needing Theorems 24 or 25, so I will omit them.

Theorem 26. *If two triangles have the two angles equal to two angles respectively, and one side equal to one side, namely, either the side adjoining the equal angles, or that subtending one of the equal angles, they will also have the remaining sides equal to the remaining sides and the remaining angle to the remaining angle.*

The two parts of the theorem have different proofs. I will restate each part separately and prove the second. Completing the proof of the first is an exercise.

The first part is attributed to Thales.

Theorem 26 (a) (ASA). *If two angles and the included side of one triangle are equal respectively to two angles and the included side of another triangle, then the triangles are congruent.*

Proof.
1. Let "ABC" and "DEF" be the triangles with,   hypothesis
   say, $\angle ABC = \angle DEF$, $BC = EF$, and $\angle ACB = \angle DFE$ (see Figure 73).
   (The idea is to show that $AB = DE$ and use SAS.)
2. Pretend $AB \neq DE$, say $AB > DE$.   RAA hypothesis
3. From AB cut off "G"$B = DE$.   Th 3
4. Draw GC.   Post 1
(Now go to it.)

I have chosen to prove AAS rather than ASA because the AAS congruence criterion is probably less familiar to you. If authors of modern textbooks

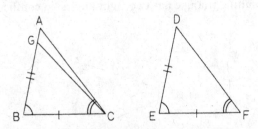

Figure 73

explicitly mention it at all, they usually wait until they have proven that the angle-sum of a triangle is 180°—Euclid's Theorem 32 (b)—because that fact, together with ASA, instantly entails AAS. AAS then appears to be a minor theorem, not worthy of much attention. However, even though the validity of AAS carries over into non-Euclidean geometry, this kind of proof does not because in non-Euclidean geometry Theorem 32 (b) is false. As students of *both* geometries we would not want to use such a proof, only to have to re-prove AAS later. Fortunately Euclid comes to our rescue by proving AAS *now*.

His proof is of course independent of Theorem 32 (b) and will remain valid in non-Euclidean geometry.

Theorem 26 (b) (AAS). *If two angles of one triangle are equal respectively to two angles of another triangle, and the sides subtending one pair of equal angles are equal, then the triangles are congruent.*

Proof.

1. Let "*ABC*" and "*DEF*" be the triangles with,   hypothesis
   say, $\angle ABC = \angle DEF$, $\angle ACB = \angle DFE$, and
   $AB = DE$ (see Figure 74).

(The idea is to show $BC = EF$, at which point the triangles will be congruent by either SAS or the just-proved ASA.)

2. Pretend $BC \neq EF$, say $BC > EF$.   RAA hypothesis
3. From BC cut off B"G" $= EF$.   Th 3
4. Draw AG.   Post 1
5. Triangles ABG and DEF are congruent.   1 ($AB = DE$,
                                                       $\angle ABC = \angle DEF$),
                                                       3, SAS
6. Therefore $\angle AGB = \angle DFE$,   Def "congruent"
7. and hence $\angle AGB = \angle ACB$.   1 ($\angle ACB = \angle DFE$).
                                                       6, C.N. 1
8. But $\angle AGB > \angle ACB$.   Th 16 ($\triangle AGC$)
9. Contradiction.   7 and 8

(So $BC \ngtr EF$. Omitting the argument that $BC \nless EF$ because it is similar, we can conclude:)

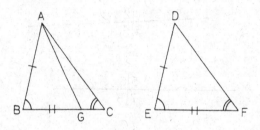

Figure 74

10. Therefore $BC = EF$. 2–9, logic
11. Therefore triangles ABC and DEF are congruent. 1 ($\angle ABC = \angle DEF$, $\angle ACB = \angle DFE$), 10, ASA or 1 ($AB = DE$, $\angle ABC = \angle DEF$), 10, SAS

Theorem 27. *If a straight line falling on two straight lines make the alternate angles equal to one another, the straight lines will be parallel to one another.*

Theorem 28. *If a straight line falling on two straight lines make the exterior angle equal to the interior and opposite angle on the same side, or the interior angles on the same side equal to two right angles, the straight lines will be parallel to one another.*

In these theorems, also in Theorem 29, Euclid uses quasi-technical language to refer to the angles formed when a straight line crosses two other straight lines. Refer to Figure 75. There are two pairs of "alternate angles": 3 and 6, 4 and 5. There are four "exterior angles," each with its corresponding "interior and opposite angle on the same side"; these pairs are 1 and 5, 2 and 6, 7 and 3, 8 and 4. Finally there are two pairs of "interior angles on the same side": 3 and 5, 4 and 6.

In the hypotheses of Theorems 27 and 28, and the conclusion of Theorem 29, Euclid's attention is on eight relationships these pairs of angles occasionally have: $\angle 3 = \angle 6$, $\angle 4 = \angle 5$; $\angle 1 = \angle 5$, $\angle 2 = \angle 6$, $\angle 7 = \angle 3$, $\angle 8 = \angle 4$; and $\angle 3 + \angle 5 = 180°$, $\angle 4 + \angle 6 = 180°$. I leave it as an exercise to prove, using Theorems 13, 15, and the Common Notions, that the following Lemma (auxiliary theorem) holds.[20]

Lemma. *If a straight line falling on two straight lines causes any one of the eight aforementioned angle relationships to occur, then it causes the other seven to occur as well.*

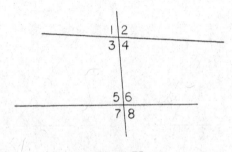

Figure 75

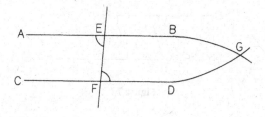

Figure 76

In view of this fact we can summarize Theorems 27 and 28 in the single statement

Theorem 27/28. *If a straight line falling on two straight lines causes any one of the eight angle relationships to occur, then the two straight lines are parallel*

—which we will prove instead.

Proof.
1. Let "*AB*" and "*CD*" be the two straight lines and   hypothesis
   "*EF*" the straight line intersecting them—at *E*
   and *F* respectively—that causes any one of the
   eight angle relationships to occur (see Figure 76).
2. Then $\angle AEF = \angle EFD$.   Lemma

(We have to show that *AB* and *CD* are parallel, i.e., by Definition 23, that no matter how much we produce them they will never intersect.)

3. Pretend that *AB* and *CD*, when produced, meet at   RAA hypothesis
   a point "*G*" which is, say, on the side of *EF* containing *B* and *D*.
4. $\angle AEF > \angle EFD$ Th 16 ($\triangle EFG$)
5. Contradiction.   2 and 4
6. Therefore *AB* and *CD*, if produced, will not meet   3–5, logic
   on the side of *EF* containing *B* and *D*.

(A similar argument shows that *AB* and *CD* will not meet on the other side of *EF*, either.)

7. Therefore *AB* and *CD* are parallel.   Def "parallel"

Theorems 29 and 30 are in the next section because they depend on Postulate 5. Theorem 31 is the only Theorem remaining in Book I that can be proven without Postulate 5; in it, having just raised the issue of parallel straight lines, Euclid verifies that such things actually exist.

Theorem 31. *Through a given point [not on a given straight line, and not on that straight line produced, it is possible] to draw a straight line parallel to [the] given straight line.*

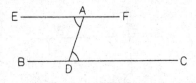

Figure 77

Proof.

1. Let "*A*" be the given point not on the given straight line "*BC*", and not on *BC* produced (see Figure 77).   hypothesis
2. Choose "*D*" at random on *BC*.   Post 10 (ii)
3. Draw *AD*.   Post 1
4. Draw *E*"*A*" making ∠*EAD* = ∠*ADC*.   Th 23
5. Produce *EA* to "*F*".   Post 2
6. *EAF* is parallel to *BC*.   4, Th 27/28

Theorems Proven With Postulate 5

The remaining theorems of Book I (29, 30, 32–48) depend on Postulate 5. Of these I have selected nine (culminating in the famous Theorem of Pythagoras) that I think will contrast nicely with the non-Euclidean geometry we will develop in Chapter 6. Since theorems that are proven without Postulate 5 remain true in that other geometry, and those requiring Postulate 5 become false, it is at this point that the two geometries diverge (Figure 78). What we've done so far belongs to both; but from now on, as we follow Euclid's path, the 19th-century revolutionaries are no longer with us.

Theorem 29. *A straight line falling on parallel straight lines makes the alternate angles equal to one another, the exterior angle equal to the interior and opposite angle, and the interior angles on the same side equal to two right angles.*

In other words, if the two straight lines on which the transversal falls are parallel, it causes all eight of the angle relationships to occur. This is the converse of Theorem 27/28. Though it is easy to confuse a theorem and its converse, I urge you in this case to be particularly careful not to.

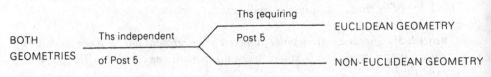

Figure 78

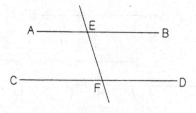

Figure 79

Proof.

1. Let "*AB*" and "*CD*" be the parallel straight lines   hypothesis
   and "*EF*" the straight line intersecting them at
   E and *F* respectively (see Figure 79).

(We will show that $\angle AEF = \angle EFD$, then invoke the Lemma on p. 84 to conclude that the other seven angle relationships occur as well.)

2. Pretend $\angle AEF \neq \angle EFD$, say $\angle AEF > \angle EFD$.   RAA hypothesis

3. $\angle AEF + \angle BEF > \angle EFD + \angle BEF$ 2, C.N. 6 (If $a > b$
                                                           then $a + c > b + c$.)

4. $\angle AEF + \angle BEF = 180°$ Th 13

5. $\angle EFD + \angle BEF < 180°$ 3, 4, C.N. 6 (If $a > b$
                                        and $a = c$ then $b < c$.)

6. *AB* and *CD* meet on the side of *EF* containing *B*   5, Post 5
   and *D*.

7. Contradiction.   1 ($AB \| CD$) and 6

(The symbol ‖ means "is parallel to.")

8. Therefore $\angle AEF = \angle EFD$.   2–7, logic

9. Therefore the other alternate angles are equal to   8, Lemma p. 84
   one another, each exterior angle is equal to the
   interior and opposite angle, and both pairs of
   interior angles on the same side are equal to two
   right angles.

Theorem 30. *Straight lines parallel to the same straight line are also parallel to one another.*

In proving this there are two cases to consider, depending on whether the straight lines are on the same or opposite sides of the straight line to which they are parallel.

Proof of the First Case.

1. Let "*AB*" and "*CD*" be straight lines parallel to   hypothesis
   the same straight line "*EF*" and lying on the
   same side of *EF* (see Figure 80).

2. On the straight line further from *EF*, say it is   Post 10 (ii)
   AB, choose a point "*G*" at random and on *EF*
   choose "*H*" at random.

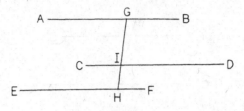

Figure 80

3. Draw *GH*.   Post 1
4. *GH* intersects *CD* (produced if necessary) in a   Post 2, Post 7 (ii)
   point "*I*".
5. ∠*BGI* = ∠*EHI*   1 (*AB*∥*EF*), Th 29
6. ∠*DIH* = ∠*EHI*   1 (*CD*∥*EF*), Th 29
7. ∠*BGI* = ∠*DIH*   5, 6, C.N. 1
8. ∠*CIG* = ∠*DIH*   Th 15
9. ∠*BGI* = ∠*CIG*   7, 8, C.N. 1
10. *AB* is parallel to *CD*.   9, Th 27/28
(The case that *AB* and *CD* are on opposite sides of *EF* is an exercise.)

The dependence of this half of the proof on Postulate 5 is in steps 5 and 6, for Postulate 5 was used to prove Theorem 29. Had I confused Theorem 29 with Theorem 27/28 (itself used in step 10) we would have been led, later on when we do non-Euclidean geometry, to the erroneous conclusion that Theorem 30 carries over.

Theorem 32. *In any triangle, if one of the sides be produced,*

(*a*) *the exterior angle is equal to the two interior and opposite angles, and*
(*b*) *the three interior angles of the triangle are equal to two right angles.*

Theorem 32 (b) is the first theorem to surprise me. Triangles don't all have the same angles, or sides, or perimeters, or areas, so why should they have the same angle-*sum*? Even looking at Figure 81 I find it hardly obvious that the three angles of triangle *ABC* have exactly the same sum as the two right angles

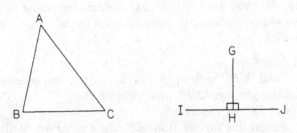

Figure 81

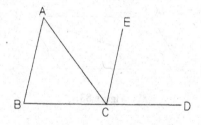

Figure 82

GHI, GHJ. My intuition is startled at the theorem's universality and exactness.

Proof.

1. Let "*ABC*" be a triangle with, say, *BC* produced   hypothesis
   to "*D*".
2. Through *C* draw *C*"*E*" parallel to *AB* (see   Th 31
   Figure 82).
   (We will show first that ∠*ACD* = ∠*BAC* + ∠*ABC*.)
3. ∠*BAC* = ∠*ACE*   Th 29 (*AC* on *AB*‖
                                                    CE)
4. ∠*ABC* = ∠*ECD*   Th 29 (*BD* on *AB*‖
                                                    CE)
5. ∠*BAC* + ∠*ABC* = ∠*ACE* + ∠*ECD*   3, 4, C.N. 2
6. ∠*ACD* = ∠*ACE* + ∠*ECD*   C.N. 4
7. ∠*ACD* = ∠*BAC* + ∠*ABC*   5, 6, C.N. 1
   (Now we will show that ∠*ACB* + ∠*BAC* + ∠*ABC* = 180°.)
8. ∠*ACB* = ∠*ACB*   C.N. 4
9. ∠*ACB* + ∠*ACD* = ∠*ACB* + ∠*BAC* +   7, 8, C.N. 2
   ∠*ABC*
10. ∠*ACB* + ∠*ACD* = 180°   Th 13
11. Therefore ∠*ACB* + ∠*BAC* + ∠*ABC* = 180°.   9, 10, C.N. 1

The dependence on Postulate 5 is in steps 3 and 4.

In Theorem 34 Euclid starts talking about parallelograms without ever having said what "parallelograms" *are*, but the notion is clear from his subsequent use of the term.

Definition 26. A *parallelogram* is a quadrilateral in which opposite sides are parallel to one another.

I leave the proof of Theorem 34 as an exercise, with a prediction that your proof will depend at least indirectly on Postulate 5.

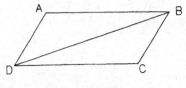

Figure 83

Theorem 34. *In [any parallelogram] the opposite sides and angles are equal to one another, and the [diagonal] bisects the area.*

Proof.
1. Let "*ABCD*" be a parallelogram with *AB*∥*DC*,   hypothesis
   AD∥*BC* in which, say, *BD* is the diagonal that has
   been drawn (see Figure 83).
(To show: *AB* = *DC*, *AD* = *BC*, ∠*A* = ∠*C*, ∠*ADC* = ∠*ABC*, and area of △*ADB* = area of △*BDC*.)

Theorems 35 through 41 are the closest Euclid ever comes to stating the familiar formulas of modern textbooks, "the area of a parallelogram is the product of its base and altitude[21]" and "the area of a triangle is half the product of its base and altitude.[21]" Of these theorems we will consider only 35, 37, and 41.

Theorem 35. *Parallelograms which are on the same base and in the same parallels are equal to one another.*

In each part of Figure 84 *ABCD* and *EFCD* are the parallelograms "on the same base" *DC* and "in the same parallels" *AF* and *DC*. For each case the theorem concludes that *ABCD* and *EFCD* "are equal to one another," i.e., that their *areas* are equal. The cases differ in the relative positions of *B* and *E*.

Proof of the First Case.
1. Let "*ABCD*" and "*EF*"*CD* be parallelograms   hypothesis
   on the same base *DC* and in the same parallels
   AF and *DC*, with *B* between *A* and *E* (see Figure 85).

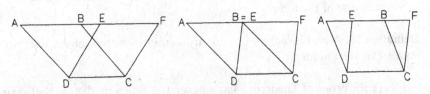

Figure 84

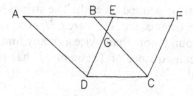

Figure 85

2. *DE* intersects *BC* in a point "*G*".   Post 7 (ii)
3. *AB = DC*   Th 34
4. *EF = DC*   Th 34
5. *AB = EF*   3, 4, C.N. 1
6. *BE = BE*   C.N. 4
7. *AE = BF*   5, 6, C.N. 2
8. *AD* is parallel to *BC*.   Def "parallelogram"
9. ∠*BAD* = ∠*FBC*   Th 29 (*AF* on
                                                      AD∥BC)
10. *AD = BC*   Th 34
11. Triangles *ADE* and *BCF* are congruent.   7, 9, 10, SAS
12. area of △*ADE* = area of △*BCF*   Post 9
13. area of △*BGE* = area of △*BGE*   C.N. 4
14. area of *ABGD* = area of *EFCG*   12, 13. C.N. 3
15. area of △*GDC* = area of △*GDC*   C.N. 4
16. Therefore, area of *ABCD* = area of *EFCD*.   14, 15, C.N. 2
(The proofs of the other two cases are exercises.)

The dependence of the first case on Postulate 5 is in steps 3, 4, 9, and 10.

Perhaps the following paradoxical situation will renew your appreciation of this theorem. Figure 86 shows two barber poles of equal heights and different

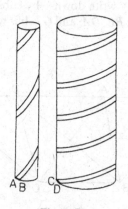

Figure 86

diameters, with one stripe painted on each. The stripes have the same width, measured horizontally (that is, arc AB = arc CD). Then, though the stripe on the stout pole goes around more than twice as many times as the stripe on the slender pole, it is a consequence of Theorem 35 that the two stripes cover exactly the same area!

Theorem 37. *Triangles which are on the same base and in the same parallels are equal to one another.*

Proof.

1. Let "ABC" and "D"BC be triangles on the   hypothesis
   same base BC and in the same parallels AD and
   $B\acute{C}$ (see Figure 87).
2. Through B draw B"E" parallel to AC; though   Th 31
   C draw C"F" parallel to BD.

(Before getting on with the proof proper we need to verify what the diagram will suggest, namely, that if BE and DA are produced they will intersect, and that if CF and AD are produced they too will intersect.)

3. Pretend that BE and DA, no matter how much   RAA hypothesis
   they are produced, will never intersect.
4. Then $BE\|DA$.   Def "parallel"
5. But $BC\|DA$.   1
6. Therefore $BE\|BC$.   4, 5, Th 30
7. But BE and BC intersect at point B.   2
8. Contradiction.   6 and 7
9. Therefore BE and DA, if produced sufficiently,   3–8, logic; Post 2
   will intersect at a point we can call "G".
10. Similarly, CF and AD, if produced sufficiently,   imitate steps 3–9
    will intersect at a point we can call "H".

(Step 10 is not, technically, a legitimate step because its reason is not one of the seven types of justification allowed (p. 12). It is rather a summary of seven steps which we could easily write down—by substituting, in steps 3–9, CF, AD, and H respectively for BE, DA, and G—but which it would be tedious to write down.)

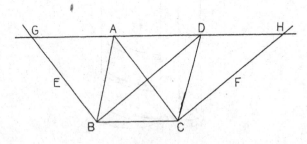

Figure 87

11. *GACB* and *DHCB* are parallelograms.	1 (*AD*∥*BC*), 2, 9, 10, Def "parallelogram"
12. area of *GACB* = area of *DHCB*	Th 35
13. ½(area of *GACB*) = ½(area of *DHCB*)	12, C.N. 6 (If $a = b$ then $\frac{1}{2}a = \frac{1}{2}b$.)
14. area of $\triangle ABC$ = ½(area of *GACB*)	Th 34
15. area of $\triangle ABC$ = ½(area of *DHCB*)	13, 14, C.N. 1
16. area of $\triangle DBC$ = ½(area of *DHCB*)	Th 34
17. Therefore, area of $\triangle ABC$ = area of $\triangle DBC$.	15, 16, C.N. 1

Theorem 41. *If a parallelogram have the same base with a triangle and be in the same parallels, the parallelogram is double of the triangle.*

Euclid's conclusion is that the *area* of the parallelogram is double the *area* of the triangle.

Proof.
1. Let "*ABCD*" be the parallelogram having the   hypothesis
   same base *DC* with the triangle "*E*"*DC*, and being
   in the same parallels *AE* and *DC* (see Figure 88).
2. Draw *AC*.   Post 1
(The proof is easy to complete using Theorems 34 and 37.)

Theorem 46. *On a given straight line [it is possible] to describe a square.*

Proof.
1. Let "*AB*" be the given straight line.   hypothesis
2. Through *A* draw a straight line perpendicular to Th 11, Post 2
   AB, and produce it to "*C*" so that *AC* > *AB* (see
   Figure 89).
3. From *AC* cut off *A*"*D*" = *AB*.   Th 3
4. Through *D* draw a straight line *D*"*E*" parallel Th 31
   to *AB*.
5. Through *B* draw a straight line *B*"*F*" parallel to Th 31
   AD.
(It is now apparent from Figure 89 that *DE* and *BF*, if produced sufficiently,

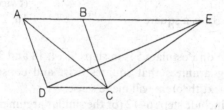

Figure 88

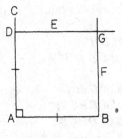

Figure 89

will intersect; but to avoid having our proof depend on the drawing, we will now *deduce* that they will.)

6. Pretend that *DE* and *BF*, no matter how much they are produced, will never intersect.	RAA hypothesis
7. Then *DE∥BF*.	Def "parallel"
8. But *AD∥BF*.	5
9. Therefore *DE∥AD*.	7, 8, Th 30
10. But *DE* and *AD* intersect at *D*.	4
11. Contradiction.	9 and 10
12. Therefore *DE* and *BF*, if produced sufficiently, will intersect at a point we can call "*G*".	6–11, logic; Post 2

(Now that we have completed our construction of the four-sided figure, we turn to verifying that it is a square.)

13. *DGBA* is a parallelogram.	4, 5, 12, Def "parallelogram"
14. Therefore *DG* = *AB* and *DA* = *GB*.	Th 34
15. But *DA* = *AB*.	3
16. Therefore the four sides of *DGBA* equal.	14, 15, C.N. 1 (several times)
17. ∠*A* = 90°	2
18. ∠*A* + ∠*D* = 180°	Th 29 (*CA* falling on *DG∥AB*)
19. ∠*D* = 90°	18, 17, C.N. 3
20. ∠*A* = ∠*G*, ∠*D* = ∠*B*	13, Th 34
21. Therefore the four angles of *DGBA* are right.	17, 19, 20, C.N. 1 (twice)
22. Therefore *DGBA* is a square.	16, 21, Def "square"

The dependence on Postulate 5 is in steps 9, 14, 18 and 20.

This theorem guarantees that squares exist and consequently that the statement of the next theorem will make sense.

Euclid doesn't include steps 6–12 (or the similar argument in steps 3–10 of the proof of Theorem 37); he accepts the testimony of the diagram. To yearn to do likewise is only natural, especially in view of the many steps involved in the alternative. After step 5 of the proof above, it must have seemed painfully obvious to you that *DE* and *BF* would meet, in fact unimaginable that they would not. I certainly wouldn't blame you if my insistence that we verify their meeting tried your patience, but I knew something you probably did not: because the verification would have to involve Postulate 5 (via Theorem 30 in step 9), in non-Euclidean geometry *DE* and *BF*, no matter how much they were produced, *would not necessarily intersect!* With the invention of an alternative geometry the intersection of straight lines has become a sensitive issue, one on which it is crucial that we keep our proofs independent of the testimony of diagrams.

Theorem 47 [Theorem of Pythagoras]. *In right-angled triangles the square on the side subtending the right angle is equal to the squares on the sides containing the right angle.*

That is, the area of the first square is the sum of the areas of the other two. A complete proof would involve so many steps that the main argument would be obscured, so here is an outline instead. Steps without reasons are exercises, the first (and most difficult) of which I've done at the end.

Proof Outline.
1. Let "*ABC*" be a right-angled triangle with, say,   hypothesis
   ∠*BAC* the right angle (see Figure 90).
2. On *BC*, *AB* and *AC* respectively construct   Th 46
   squares *BC*"*DE*", *AB* "*FG*", and *AC*"*HI*".
3. Through *A* draw *A*"*J*" parallel to *BE*.   Th 31
4. *AJ*, if produced sufficiently, will intersect *BC* in
   a point "*K*".

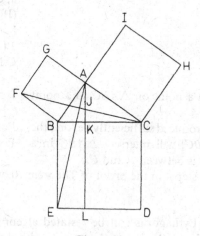

Figure 90

5. AK, if produced sufficiently, will intersect ED in a point "L".

6. Draw AE and FC.   Post 1

7. Triangles FBC and ABE are congruent.

8. area of $\triangle FBC$ = area of $\triangle ABE$ Post 9

9. 2 (area of $\triangle FBC$) = 2 (area of $\triangle ABE$)   C.N. 6 (If $a = b$ then $2a = 2b$.)

10. $ABFG$ is a parallelogram.

11. area of $ABFG$ = 2 (area of $\triangle FBC$)   Th 41

12. area of $ABFG$ = 2 (area of $\triangle ABE$)   9, 11, C.N. 1

13. $BKLE$ is a parallelogram.

14. area of $BKLE$ = 2 (area of $\triangle ABE$)   Th 41

15. area of $ABFG$ = area of $BKLE$ 12, 14, C.N. 1

16. Similarly by drawing AD and BH it follows that area of $ACHI$ = area of $CDLK$.

17. area of $ABFG$ + area of $ACHI$ = area of $BKLE$ + area of $CDLK$ 15, 16, C.N. 2

18. area of $BCDE$ = area of $BKLE$ + area of $CDLK$ C.N. 4

19. Therefore, area of $BCDE$ = area of $ABFG$ + area of $ACHI$.   18, 17, C.N. 1

Proof of Step 4.

a. $\angle JAB + \angle ABE = 180°$ Th 29 (AB on $AJ \| BE$)

b. $\angle CBE = 90°$ 2, Def "square"

c. $\angle ABE > \angle CBE$ C.N. 5

d. $\angle ABE > 90°$ b, c, C.N. 6 (If $a = b$ and $c > a$ then $c > b$.)

e. $\angle JAB < 90°$ a, d, C.N. 6 (If $a + b = c$ and $b > \frac{1}{2}c$ then $a < \frac{1}{2}c$)

f. $\angle JAB < \angle BAC$ 1 ($\angle BAC = 90°$), e, C.N. 6 (If $a = b$ and $c < b$ then $c < a$.)

g. AJ is drawn from a point on $\triangle ABC$ to a point inside $\triangle ABC$.   f

h. Therefore AJ, if produced sufficiently beyond the point inside $\triangle ABC$, will intersect $\triangle ABC$ in a point "K", which is between B and C.   Post 2, Post 6 (iii), Post 1

(The last step is like step 4 in the proof of Theorem 10, which I discussed on pp. 65–66.)

The Theorem of Pythagoras can be restated algebraically: "If the sides of a right-angled triangle have lengths a, b, c, with c subtending the right angle, then $c^2 = a^2 + b^2$." You are probably more familiar with it in this form.

To me the Theorem of Pythagoras is very surprising. Although the man-made world is filled with right angles, I think of them as originally natural events akin to lightning and the Big Dipper. I stand in a field at right angles to the ground. Facing east, I must turn through a right angle to face south. A falling acorn traces a path at right angles to the horizon. "$c^2 = a^2 + b^2$," on the other hand, evokes no visceral memories whatever. Numbers are not part of nature; and if they were, it is unlikely I would chance upon three that were so related. Because the equation is abstract and precise, it is alien. I can't imagine what such a thing could possibly have to do with everyday right angles. So, when the pall of familiarity lifts, as it occasionally does, and I see the Theorem of Pythagoras afresh, I am flabbergasted.

Index to Euclidean Geometry

Primitive Terms

Defined Terms

Axioms

Theorems Without Postulate 5

Theorems With Postulate 5

Exercises

See also Chapter 4, Exercises 1–3.

1. Prove that parallel straight lines are always the same distance apart. That is (Figure 91), let *AB* and *CD* be parallel straight lines, with *E* and *F* any two points on *AB*, and *EG* and *FH* drawn at right angles to *CD* (produced if necessary); show that *EG = FH*.

2. Well? (Figure 92.)

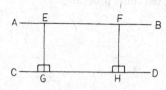

Figure 91

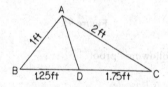

Figure 92

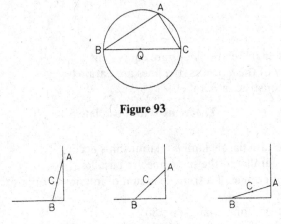

Figure 93

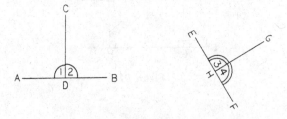

Figure 94

3. Prove that an angle inscribed in a semicircle is right. That is, in Figure 93, where Q is the center of the circle, show that $\angle BAC = 90°$.

4. In Figure 94 AB is a ladder, originally flush against the wall, whose foot B is being pulled to the left. Find the path of C, the ladder's midpoint, and prove your answer is correct.

5. Standing at the edge of the sea you see a small boat offshore. You have a compass (a normal one that stays open), a straight stick a little shorter than you are, a few inches of adhesive tape, and a friend. You know the length of your stride and the ASA theorem. How could you find the distance to the boat? (The attribution of the ASA theorem to Thales is based on the legend that he knew how to do this.)

6. Given, in Figure 95, that $\angle 1 = \angle 2$ and $\angle 3 = \angle 4$, Definition 10 tells us that each of angles 1, 2, 3, and 4 is right. Pick up $GEHF$ and set it down on $CADB$ so that EF runs along AB and H coincides with D. Prove, *without* using Postulate 4, that $\angle 1 = \angle 3$. Thus, if we were to allow superposition of figures, we would not need to assume Postulate 4, but could prove it.

Figure 95

7. Find the flaw in the following "proof."

"Theorem." *Every triangle is isosceles.*

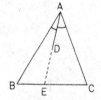

Figure 96

"Proof."

1. Let *"ABC"* be a triangle (see Figure 96).   hypothesis
(To show: △*ABC* is isosceles.)
2. Draw *A"D"* to bisect ∠ *BAC*.   Th 9
3. Produce *AD* until it intersects the triangle one more   Post 2, Post 6 (iii)
   time, at a point *"E"*.
4. *E* is not on *AB* or *AC*.   Post 1
5. Therefore *E* is on *BC* (as drawn).   3, 4

Case 1. *AE* is at right angles to *BC* (see Figure 97).
6. Then triangles *ABE* and *ACE* are congruent.   ASA
7. Therefore *AB = AC*, and   Def "congruent"
8. △*ABC* is isosceles.   Def "isosceles"

Case 2. *AE* is not at right angles to *BC*; say, to be specific, ∠ *AEC* < 90° (see Figure 98).
(The argument would be the same if ∠ *AEB* were the angle less than 90°.)
9. Bisect *BC* at *"F"*.   Th 10
10. Draw *F"G"* at right angles to *BC*.   Th 11

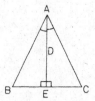

Figure 97

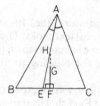

Figure 98

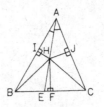

Figure 99

11. $\angle AEC + \angle GFE < 180°$	hypothesis of Case 2, 10, C.N. 6 (If $a < b$ and $c = b$ then $a + c < 2b$.)
12. *EA* and *FG*, if produced sufficiently, will meet.	Post 5
13. Produce *FG* until it meets *EA* at "*H*".	Post 2
14. Draw *H"I"* perpendicular to *AB* and *H"J"* perpendicular to *AC* (see Figure 99).	Th 12
15. Draw *HB* and *HC*.	Post 1
16. Triangles *HBF* and *HCF* are congruent.	9, 10, SAS
17. Therefore *HB* = *HC*.	Def "congruent"
18. Triangles *AIH* and *AJH* are congruent.	AAS
19. Therefore *HI* = *HJ*.	18, Def "congruent"
20. $HB^2 = HC^2$	17, C.N. 6 (If $a = b$ then $a^2 = b^2$.)
21. $HI^2 = HJ^2$	19, C.N. 6 (Same.)
22. Therefore $HB^2 - HI^2 = HC^2 - HJ^2$.	20, 21, C.N. 3
23. But $HB^2 = HI^2 + IB^2$,	Th of Pythagoras
24. so $HB^2 - HI^2 = IB^2$;	C.N. 3
25. and $HC^2 = HJ^2 + JC^2$,	Th of Pythagoras
26. so $HC^2 - HJ^2 = JC^2$;	C.N. 3
27. therefore $IB^2 = JC^2$,	22, 24, 26, C.N. 1 (twice)
28. and *IB* = *JC*.	C.N. 6 (If $a^2 = b^2$ and a and b are positive then $a = b$.)
29. But *AI* = *AJ*.	18, Def "congruent"
30. Therefore *AB* = *AC*, and	29, 28, C.N. 2
31. △*ABC* is isosceles.	Def "isosceles"

Notes

[1] *Theon of Alexandria* was a professor at the Museum who edited Euclid's works and wrote commentaries on astronomical textbooks. Theon's daughter Hypatia is the first woman mathematician whose name has come down to us. In 415 she was torn to pieces by a mob of Christian fanatics for refusing to abandon her pursuit of "pagan" learning.

[2] *Sir Thomas L. Heath*, 1861–1940, was an expert on Greek mathematics who translated some of its finest works into English. His translation of the *Elements* is described

in the Bibliography. Volume 1 of his *A History of Greek Mathematics* (1921; Dover reprint, 1981) contains a detailed account of the transmission of the *Elements*. (Henceforth, all quotations of Euclid, De Morgan, Pappus and Heath are from Sir Thomas L. Heath: *The Thirteen Books of Euclid's Elements*, 1908, 1925; Dover reprint, 3 volumes, 1956. Reprinted with permission of Cambridge University Press.)

[3] *Einstein once said* Reported in Lincoln Barnett's *The Universe and Dr. Einstein* (William Sloane Associates, 1948), p. 52.

[4] *as numerous as the points of a line segment.* I'm lying here, because the whole truth would complicate the discussion enormously. It is *traditional* mathematics that cannot combine so many quantities. If you happen to have heard of the new technique called "nonstandard analysis," read the next note.

[5] *the two alternatives.* This note won't make sense unless you have heard of the new mathematical technique called "nonstandard analysis."

With the invention of nonstandard analysis, it is now possible to "add" uncountably many zeroes, and the sum is—of course—zero. This strengthens the argument against Euclid's Definition 1. But *between* the two alternatives there is now another: that points have an infinitesimal diameter dx, which is greater than zero but less than every standard positive number. The sum of the lengths of the points in our 1-meter segment would then be the nonstandard number $\sum_0^1 dx$, which determines the unique standard number $\int_0^1 dx = 1$.

This would have delighted the Greeks. They had thought long and hard about infinitesimals—Book III, Theorem 16 of the *Elements*, for example, amounts to a proof that the magnitude of the "angle" between a circle and tangent is infinitesimal. But as they could contrive no consistent theory of magnitudes that would include such things, they banished them by contriving instead the Axiom of Archimedes (appearing in the *Elements* as Definition 4 of Book V):

If a and b are any two magnitudes of the same kind and $a < b$, then there exists a positive whole number n such that $2^n \cdot a > b$.

[6] *what the standard of mathematical reasoning implies.* "It is well known," runs a comment added to Book X of the *Elements* around A.D. 450, "that the man who first made public the theory of irrationals"—this would be Hippasos of Metapontion (p. 3), who discovered that $\sqrt{2}$ is not rational—"perished in a shipwreck, in order that the inexpressible and unimaginable should ever remain veiled. And so the guilty man, who fortuitously touched on and revealed this aspect of living things, was taken to the place where he began and there is forever beaten by the waves." (Carruccio, *op. cit.*, p. 27.)

This comment's unmistakable tone of moral outrage reflects, I think, the commentator's appreciation of the real significance of Hippasos's discovery, namely, that *the mind can be divided against itself.* What Hippasos's discovery signaled, and the standard of mathematical reasoning acknowledges, is humanity's loss of intellectual innocence.

[7] *Archimedes* of Syracuse, 287–212 B.C., was the greatest mathematician of antiquity. Beside being the person who called a straight line "the shortest path between two points," it was Archimedes who

first devised a method for calculating π to any degree of accuracy,
found the volume of a sphere,
invented a method of integration,
ran home naked from a public bath shouting "Eureka!",
said, "Give me a place to stand and I can move the earth," and
devised war machines so ingenious that they held a besieging Roman army off Syracuse for three years.

[8] "... *is the rigor thereof*" Maxim of the American mathematician Robert L. Moore, 1882–1974, inventor of the "Moore method" for training mathematicians. "While he was still a graduate student," writes F. Burton Jones in the *American Mathematical Monthly* (April, 1977, p. 274),

... Moore conceived the basic idea that led eventually to his radical method of teaching. With his quick mind and restless spirit he found the lecture method rather boring—in fact, mind-dulling. To liven up a lecture he would run a race with his professor by seeing if he could discover the proof of an announced theorem before the lecturer had finished his presentation. Quite frequently he won the race. But in any case, he felt that he was better off for having made the attempt. So if one could get students to prove the theorems for themselves, not only would they have a deeper and longer-lasting understanding, but somehow their ability and interest would be strengthened.

[9] *its admission would lead to absurdities like 1 = 2*. Two rules of the number system are

(1) $0 \cdot x = 0$,   and

(2) $x \cdot \dfrac{1}{x} = 1$,   whenever $\dfrac{1}{x}$ is a number.

If the combination of symbols "1/0" were allowed to be a number, then

$$0 \cdot \frac{1}{0} = 0$$

by rule (1), and

$$0 \cdot \frac{1}{0} = 1$$

by rule (2), making $0 = 1$. That $1 = 2$ would then follow by adding 1 to both sides.

[10] *when equals are added to unequals, the sums are unequal*. Interestingly, most manuscripts of the *Elements* include precisely this statement among the Common Notions, but it appears to have been interpolated by Theon of Alexandria (p. 22).

[11] *a prime number* is a whole number, greater than 1, that is divisible only by itself and 1. The first ten prime numbers are 2, 3, 5, 7, 11, 13, 17, 19, 23, and 29. There are technical reasons for not counting 1 as a prime.

[12] *Postulate 6*. In another branch of mathematics Postulate 6 (i) is known as (a special case of) the "Jordan Curve Theorem" and (ii) is a corollary. In geometry the triangular case of (iii) is called "Pasch's Axiom" after Moritz Pasch, 1843–1930, a pioneer in the work of ferreting out Euclid's unstated assumptions. It was Pasch who formulated the "Pattern for a Material Axiomatic System" (p. 6).

[13] *Augustus De Morgan*, 1806–1871, was an English mathematician who formulated two "laws" of set theory you may remember from school. If A and B are sets, $\mathcal{U}$ is the universal set (this idea is itself due to De Morgan), "∪" and "∩" denote union and intersection, and $\mathcal{U}$ "minus" a set indicates the set of all things in $\mathcal{U}$ *not* in the set, then "De Morgan's Laws" are that

$$\mathcal{U} - (A \cup B) = (\mathcal{U} - A) \cap (\mathcal{U} - B)$$

and

$$\mathcal{U} - (A \cap B) = (\mathcal{U} - A) \cup (\mathcal{U} - B).$$

[14] *we have no control over the direction in which AH is drawn*. Except to draw $\triangle ABD$ with D under AB, which gives us one alternative, or to begin the construction by drawing AC instead of AB, which gives us two more.

[15] *that figures may be moved*. Also, there are inadequate reasons for steps 3–5. For example in step 3, Euclid is saying that since AB and DE are equal, they coincide. But

no Common Notion says, "Things that are equal (if superposed, will) coincide"; Common Notion 4 makes the *converse* assertion, but that's not the same thing. (The "converse" of a conditional statement is obtained by interchanging the hypothesis and conclusion. Though easily confused, a statement and its converse say different things. For example the converse of, "If one is a woman then one is a human being," is, "If one is a human being then one is a woman.")

[16] *as little attention to it as possible.* Undermining my opinion is the fact that Euclid's other use of superposition, in Theorem 8, can be easily avoided. "Since the alternative proof of Theorem 8 is not very difficult," writes Mueller in *Philosophy of Mathematics and Deductive Structure in Euclid's Elements* (MIT, 1981), "it seems likely that if Euclid did wish to avoid superposition, his wish was not deep-seated enough to cause him to search very hard for alternatives to it."

[17] *René Descartes*, philosopher, scientist, and mathematician, 1596–1650, is remembered best for applying mathematical methods to philosophy. He began by questioning everything, and found that the only thing he was unable to doubt was the very fact of his doubting. "I think," he said, and concluded, "therefore I am." (*Cogito, ergo sum.*) On this theorem he constructed his philosophical system.

[18] *Moritz Pasch.* See the "Postulate 6" note on p. 104.

[19] *Supplementary* is a term frequently used in modern textbooks to describe angles related as angles 1 and 2 in Figure 49. Formally, "supplementary" angles are angles having a common vertex and side, the other sides of which are on opposite sides of the common side and lie in a straight line with one another.

[20] *the following Lemma holds.* As a complete proof of the Lemma would involve $8 \times 7 = 56$ individual demonstrations, I'm not seriously suggesting you do that. If you were to choose a representative assortment of 3 or 4 cases and prove them, you'd be satisfied that the others could be proven similarly.

[21] *altitude.* An "altitude" of a parallelogram is the perpendicular distance between a base and the opposite side. An "altitude" of a triangle is the perpendicular distance between a base and the opposite vertex.

CHAPTER 3

Geometry and the Diamond Theory of Truth

Reviewing the last chapter I am reminded of how awed I was by geometry when I studied it in high school, and of how that feeling deepened when, years later, I read the *Elements* itself. Based on what seemed indubitable principles, buttressed by what I found to be impeccable logic, Euclid's edifice loomed in my consciousness as a marvel among sciences, unique in its clarity and unquestionable validity.

For more than two millenia the *Elements* has had this sort of effect on its readers, especially those of a philosophic bent. The physiologist, physicist, and philosopher Hermann von Helmholtz (1821–1894) described his experience in the opening of a lecture delivered in 1870:

The fact that a science can exist and can be developed as has been the case with Euclidean geometry, has always attracted the closest attention among those who are interested in questions relating to the bases of the theory of cognition. Of all branches of human knowledge, there is none which, like it, has sprung as a completely armed Minerva from the head of Jupiter; none before whose death-dealing Aegis doubt and inconsistency have so little dared raise their eyes. It escapes the tedious and trouble-some task of collecting experimental facts, which is the province of the natural sciences in the strict sense of the word; the sole form if its scientific method is deduction. Conclusion is deduced from conclusion, and yet no one of common sense doubts but that these geometrical principles must find their practical application in the real world about us. Land surveying, as well as architecture, the construction of machinery no less than mathematical physics, are continually calculating relations of space of the most varied kind by geometrical principles; they expect that the success of their construc-tions and experiments shall agree with these calculations; and no case is known in which this expectation has been falsified, provided the calculations were made cor-rectly and with sufficient data.*

In view of this admiration the *Elements* has consistently elicited over the years, and of the prestigious role it concomitantly maintained until the mid-19th century as scientific archetype, it is not surprising that we find in the

---

* Reprinted in Hermann von Helmholtz: *Popular Scientific Lectures*, Morris Kline, ed., (p. 223) Dover Publications, Inc., New York, 1962. Reprinted with permission.

history of philosophy a concept of truth sustained by the example of the *Elements*, whose influence in philosophy runs parallel to that of the *Elements* in science. This concept I will call the "Diamond Theory" of truth.

Kant's Distinctions

Nineteenth-century discussions of the nature of geometric truth were dominated by the doctrine of Immanuel Kant[1] (1724–1804). Kant's philosophy embodies the Diamond Theory in highly refined form.

Reflecting on the work of earlier philosophers, Kant drew a distinction between statements he called "analytic" and those he called "synthetic." (Though allied to, this distinction is different from the one we drew on p. 58 between "analytic" and "synthetic" *demonstrations*.) Examining this distinction, together with an older one Kant inherited—between "a priori[2]" and "empirical" statements—will direct our attention to the issues with which the Diamond Theory is involved.

I'll let Kant speak for himself, beginning with the distinction he inherited.

It is ... a question which deserves ... investigation, and cannot be disposed of at first sight, whether there exists a knowledge independent of experience, and even of all impressions of the senses? Such knowledge is called *a priori*, and distinguished from *empirical* knowledge, which has its sources ... in experience.

This term *a priori*, however, is not yet definite enough to indicate the full meaning of our question. For people are wont to say, even with regard to knowledge derived from experience, that we have it, or might have it, *a priori*, because we derive it from experience, not *immediately*, but from a general rule, which, however, has itself been derived from experience. Thus one would say of a person who undermines the foundations of his house, that he might have known *a priori* that it would tumble down, that is, that he need not wait for the experience of its really tumbling down. But still he could not know this entirely *a priori*, because he had first to learn from experience that bodies are heavy, and will fall when their supports are taken away.

We shall therefore ... understand by knowledge *a priori* knowledge which is *absolutely* independent of all experience, and not of this or that experience only. Opposed to this is empirical knowledge, or such as is possible ... only ... by experience.

(Critique of Pure Reason, Introduction, I*)

Kant does not count as "experience" what I shall call "linguistic" experience, namely, whatever we must undergo to understand our language's words and sentences and to analyze them logically. He would say for example that the statement, "All bachelors are unmarried,[3]" is known a priori because I require no experience other than language-learning to judge it true. On the other hand he would say that the statement, "Of the bachelors who are presently members of the Stonehill College faculty, exactly two were born in 1946," assuming I know it to be true, expresses empirical knowledge because in order to judge it true I must have gone beyond my understanding of the

---

* Translated by Max Müller, 2nd edition, Macmillan, London, 1927.

statement itself, and beyond logical deductions from it—by going to the College and personally interviewing its faculty, perhaps, or talking to someone who has, or studying the payroll records, etc. Of course we would expect "All bachelors are unmarried" to be confirmed by these extra-linguistic experiences as well, but the point is that a priori statements (I prefer to speak of "statements" rather than "knowledge") do not *need* such validation while empirical statements *do*.

So a statement we judge to be true is "a priori" if it is justifiable without extra-linguistic experience, and "empirical" if its justification requires extra-linguistic experience. The distinction seems clear enough. You may find it blur, however, as Kant moves on to his new distinction, for he draws it hard by the old one.

In all judgments in which there is a relation between subject and predicate (I speak of affirmative judgments only, the application to negative ones being easy), that relation can be of two kinds. Either the predicate B belongs to the subject A as something contained (though covertly) in the concept A; or B lies outside the sphere of the concept A, though somehow connected with it. In the former case I call the judgment analytic, in the latter synthetic .... [I]n the former nothing is added by the predicate to the concept of the subject, but the concept is only divided into its constituent concepts which were always conceived as existing within it, though confusedly; while the latter adds to the concept of the subject a predicate not conceived as existing within it, and not extracted from it by any process of mere analysis.

It becomes clear from this,

1. That our knowledge is in no way extended by analytic judgments, but that all they effect is to put the concepts which we possess into better order and render them more intelligible.
2. That in synthetic judgments I must have besides the concept of the subject something else (X) on which the understanding relies in order to know that a predicate, not contained in the concept, nevertheless belongs to it.

(*Critique of Pure Reason*, Introduction, IV)

This can be paraphrased as follows. A statement (I prefer to apply Kant's terms to "statements" rather than "judgments") we judge to be true is "analytic" if mere understanding of what it says (including logical analysis of what it says) is sufficient to support our judgment, whereas it is "synthetic" if, in addition to understanding, we require something else—Kant's "X."

We can use the same two examples as before. "All bachelors are unmarried" is an analytic statement because in judging it true nothing more than my understanding of its meaning is required. But "Of the bachelors who are presently members of the Stonehill College faculty, exactly two were born in 1946" is synthetic because in addition to understanding it I require something else (a survey of the faculty, an informant, a look at the College's records) before I can conclude it is true.

It's hard to see a difference between a priori and analytic statements because

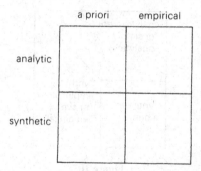

Figure 100

any statement we are likely to think of as being an example of one will be an example of the other as well. Our example, "All bachelors are unmarried," is both a priori *and* analytic. Similarly it's hard to see a difference between empirical and synthetic statements: "Of the bachelors who are presently members of the Stonehill College faculty, exactly two were born in 1946" is both empirical *and* synthetic. Nonetheless the two distinctions are not identical, at least in theory.

In Figure 100 the large square represents the collection of all statements that I judge to be true. The vertical line separates the a priori statements from those that are empirical. It's clear that this division is sharp, that is that the two columns do not overlap, because the terms are simply negations of one another—empirical statements are the statements that require extra-linguistic experience for justification, a priori statements are the statements that do not. The horizontal line represents the boundary between statements that are analytic and those that are synthetic. Again the division is sharp, because "analytic" means simply "not synthetic"—synthetic statements are the statements that require, beyond understanding and logical analysis, an extra factor X for justification, analytic statements are the statements that do not.

When the two distinctions are superimposed, as they are in Figure 100, the top right quadrant is really empty. No statement can be both analytic *and* empirical, both justified *merely* by my understanding and analysis of what it says and at the same time justified only by some experience *beyond* what I require to understand and analyze it. Thus we should revise our drawing into Figure 101. (Another way to see why the top right quadrant of Figure 100 is empty is to note that every analytic statement is automatically a priori and every empirical statement automatically synthetic.)

Also, it is clear that the two sectors labeled "analytic a priori" and "synthetic empirical" in our new figure are *not* empty, as we have considered an example of each; indeed, it was our feeling that every statement is either both analytic *and* a priori, or both synthetic *and* empirical, that moved us to consider all these combinations in the first place.

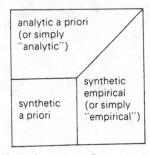

Figure 101

What is not clear is whether the corner of Figure 101 labeled "synthetic a priori" is empty. "A priori" means "verifiable without recourse to experience beyond what I need to understand and analyze" and "synthetic" means "verifiable only by something else (*X*) *in addition* to understanding and analysis." If you feel the "something else," the *X*, could only be some extra-linguistic *experience*, then it follows that the terms are incompatible and the "synthetic a priori" corner empty. But while I myself have difficulty imagining what else Kant's *X* could be, I am forced to admit—since by "experience" Kant means information that comes to us through our *senses*—there remains the logical possibility of some *nonsensory* factor whereby a synthesis could be obtained. We would thus have a statement at once synthetic, because of the extra factor, and a priori because that nonsensory extra factor would not be what Kant calls "experience." Such an *X* might be some sort of inner perception or insight.

Synthetic A Priori Statements

Kant was interested in synthetic a priori statements because, were such statements to exist, they would be the best kind.

Kant felt that empirical statements, because of their reliance on information coming to us through our senses, are necessarily uncertain. Consider our example: "Of the bachelors who are presently members of the Stonehill College faculty, exactly two were born in 1946." Say I believe this to be true—that I have examined the College's records and confirmed my findings by interviewing the faculty's bachelors. There remains ample room for doubt. Maybe someone's file contains a clerical error, and coincidentally I misheard what the same person said to me. Maybe a "bachelor" is secretly married, or some of the faculty lie about their ages. Maybe a bachelor who was actually born on December 14, 1945 *thinks* he was born on January 5, 1946, because the people he believes to have been his biological parents (who died before they could tell him he was adopted) took him into their home at the apparent age of one year on January 5, 1947 and so designated January 5, 1946 as his date of birth! I have to agree with Kant. I can't imagine an empirical statement that cannot be doubted.

Kant also felt that, conversely, a priori statements *are* certain, but that the bulk of them—the analytic statements—are insubstantial. "[O]ur knowledge is in no way extended" by analytic statements, he said, because "all they effect is to put the concepts which we possess into better order and render them more intelligible." It is doubtless true that "All bachelors are unmarried," but it is also trivial.

Thus of the three kinds of statements we judge to be true (Figure 101), two stand condemned—empirical statements because they are uncertain, analytic statements because they are uninformative.

A synthetic a priori statement, Kant felt, would have neither imperfection. Being synthetic it would be more than an elaboration of something already known implicitly; it would provide information that is genuinely *new*. And being a priori it would have the certainty Kant felt to be inherent in any knowledge independent of the testimony of our senses.

Geometry as Synthetic A Priori

Kant found the theorems of Euclidean geometry to be synthetic a priori statements.

His finding that they were a priori was nothing new. That had been the majority opinion among scholars for at least 2200 years, but for completeness he reviewed its basis anyway. (Henceforth all Kant quotations are from *Prolegomena to Any Future Metaphysics*, Preamble, §2.*)

First of all, we must observe that all strictly mathematical judgments are *a priori*, and not empirical, because they carry with them necessity, which cannot be obtained from experience. But if this not be conceded to me, very good; I shall confine my assertion to *pure mathematics*, the very notion of which implies that it contains pure *a priori* and not empirical knowledge.

Kant's contribution, made possible by the analytic/synthetic distinction, was the finding that the theorems of Euclidean geometry are also synthetic.

This fact seems hitherto to have altogether escaped the observation of those who have analyzed human reason; it even seems directly opposed to all their conjectures ....

According to Kant, "their conjectures" had been that geometry consisted of nothing but analytic statements.

---

* Reprinted with permission of Macmillan Publishing Company from Immanuel Kant, *Prolegomena to Any Future Metaphysics*, edited by Lewis White Beck. Copyright © 1950 by Macmillan Publishing Company, renewed 1978 by Lewis White Beck.

For as it was found that the conclusions of mathematicians all proceed according to the law[s] of [logic] ... men persuaded themselves that the fundamental principles were known from the same law[s]. This was a great mistake, for a synthetic proposition can indeed be established by the law[s] of [logic], but only by presupposing another synthetic proposition from which it follows ....

Thus, if Kant could show that Euclid's "fundamental principles" are synthetic, his thesis that the theorems are synthetic would follow. His position was that the Postulates, but not the Common Notions, are synthetic.

Some ... principles, assumed by geometers, are indeed actually analytic ...; but they only serve, as identical propositions, as a method of concatenation, and not as principles—for example $a = a$, the whole is equal to itself, or $a + b > a$, the whole is greater than its part.

When Kant argued that the Postulates, on the other hand, are synthetic, he took as his example "A straight line is the shortest path between two points." This is not, of course, one of Euclid's Postulates (see p. 31), but geometry textbooks have frequently contained it nonetheless. Kant was simply quoting the book he had used in school.

That a straight line is the shortest path between two points is a synthetic proposition. For my concept of straight contains nothing of quantity, but only a quality. The concept "shortest" is therefore altogether additional and cannot be obtained by any analysis of the concept "straight line." ... What usually makes us believe that the predicate of such apodictic [=incontestable] judgments is already contained in our concept, and that the judgment is therefore analytic, is the duplicity of the expression. We must think a certain predicate as attached to a given concept, and necessity indeed belongs to the concepts. But the question is not what we must join in thought *to* the given concept, but what we actually think together with and in it, though obscurely ....

This is the only example Kant gave. He claimed that on examination *every* Postulate would likewise be revealed as synthetic. It is unfortunate for us, as students of Euclid's original system, that Kant's crucial example is something we don't acknowledge to be a Postulate.

Let's construct our own example, along Kantian lines. For this purpose Postulate 4—"All right angles are equal"—is a good text. The question is, does mere understanding of the concept "right angle," and logical analysis of that concept, lead us to necessarily predicate of right angles that they are all equal? If the answer is "yes," Postulate 4 is analytic; if "no," it is synthetic. But this question is precisely what occupied us when Postulate 4 was introduced (pp. 40–41), and we concluded then that the answer is "no." Thus Postulate 4 is synthetic.

In summary, Kant's argument that the theorems of geometry are synthetic a priori statements goes like this.

1. Euclid's Postulates, Common Notions, and Theorems are all a priori. (As we are confident that they are true, and no experimental test would increase our confidence, our judgment that they are true must not depend on extra-linguistic experience.)

2. Euclid's Postulates are also synthetic. (This has been verified for at least one Postulate.)
3. But logical consequences of synthetic statements are synthetic, and
4. every Theorem depends on the Postulates. (None is a consequence of the Common Notions alone.)
5. Therefore Euclid's Theorems are all synthetic.
6. Therefore Euclid's Theorems are synthetic a priori statements. (By 1 and 5.)

Kant's Doctrine of Space

Three sections ago we said (p. 110) that a synthetic a priori statement, if such a thing were to exist, would—because it is synthetic—require as verification something else (Kant's X) beyond understanding and logical analysis, but that this X—because the statement is a priori—could not be experience. Now that Kant has concluded that synthetic a priori statements do, indeed, exist in geometry, he is faced with the problem: For geometry, what could this extraordinary X be?

Kant's answer is his doctrine of space. To examine this in any detail would be to depart from the path of this chapter. But I want to sketch the main idea, so as not to leave you hanging.

Kant decided that data from our senses is "processed" by our unconscious minds before we receive it in consciousness. We never hear raw sound, for example. Our *ears* detect it, but this data is first transmitted to a "Sense-Data Processor" in the brain which *organizes* the data according to certain principles before passing it on to our awareness. What we "hear" (become aware of) is not real sound, then, but processed sound.

This can be applied to our notion of space. Data about the real space in which we live comes through our senses, chiefly sight and touch. But before we become aware of this data it has been processed by the Sense-Data Processor. Our notion of space, then, is actually not of real space but of processed space. It is processed space we study in geometry.

When we think about space, we find Euclid's Postulates to be true. We cannot conceive of how they could *not* be true. We are unable to doubt them. Usually, when we are unable to doubt a statement, it is because the statement is analytic; yet Euclid's Postulates contain information not extractable from their components by understanding or logical analysis: they are synthetic. How can this be?

The only explanation, declared Kant, is that Euclid's Postulates describe how the Sense-Data Processor organizes data about real space. Processed space, the space studied in geometry, is pervaded by Euclid's Postulates because they are the very principles by which it is organized! Our inability to doubt Euclid's Postulates is but a reflection of the fact that our brains are so constructed that we are literally *unable* to think about space in any other way.

For geometry, then, the X is a structure within our brains that controls the way we perceive space. And the feeling of helplessness that comes over us when we try to conceive of space in a way that would render any of Euclid's Postulates false is but our subliminal awareness of that structure.

The Diamond Theory of Truth

People have always longed for truths about the world—not logical truths, for all their utility; or even probable truths, without which daily life would be impossible; but informative, certain truths, the only "truths" strictly worthy of the name. Such truths I will call "diamonds[4]"; they are highly desirable but hard to find.

From Plato to Kant—in round numbers, from 400 B.C. to A.D. 1800—Western philosophers and scientists were, for the most part, diamond-hunters. To be sure, an occasional skeptic would scoff at the search, claiming that truths simultaneously informative and certain could not exist. But most thinkers were too busy unearthing diamonds to pay much attention; and those who did were usually able to satisfy themselves that the skeptic was mistaken.

Kant was a diamond-hunter. Synthetic a priori statements were his diamonds.

One of Kant's motives for writing the books from which I have been quoting was to disagree with the Scottish skeptic David Hume (1711–1776) about a metaphysical principle known as the Law of Causality ("Every event has a cause"). Hume had denied it was a diamond, and had done so in such a way as to threaten the existence of all diamonds.

Hume, being prompted to cast his eye over the whole field of *a priori* cognitions in which human understanding claims such mighty possessions..., heedlessly severed from it a whole ... province, namely, pure mathematics; for he imagined that its nature or, so to speak, the state constitution of this empire, depended on totally different principles, namely, on the law[s] of [logic] alone; and although he did not divide judgments ... formally and universally as I have done here [into analytic and synthetic], what he said was equivalent to this: that mathematics contains only analytic, but metaphysics synthetic, *a priori* propositions. In this, however, he was greatly mistaken, and the mistake had a decidedly injurious effect upon his whole conception. But for this, he would have extended his question concerning the origin of our synthetic judgments far beyond the metaphysical concept of causality and included in it the possibility of mathematics *a priori* also .... And then he could not have based ... metaphysical propositions on mere experience [Hume had said the Law of Causality is empirical] without subjecting the axioms of mathematics equally to experience, a thing which he was far too acute to do. The good company into which metaphysics would thus have been brought would have saved it from the danger of contemptuous ill-treatment, for the thrust intended for it must have reached mathematics, which was not and could not have been Hume's intention.

Note the crucial role played in Kant's thought by mathematics. He is sure that Hume would have accorded diamond status to the theorems of geometry. Therefore, Kant reasoned, had Hume but noticed that they were synthetic, he would never have come to the conclusion that synthetic statements are always empirical, which is what had prevented him from recognizing that the non-mathematical Law of Causality was a diamond as well. To Kant the diamond-nature of geometry was particularly apparent, and led to the discovery of other diamonds beyond the borders of mathematics.

Among diamond-hunters Kant was very conservative. His terms were so sharply defined, his analysis so painstaking, that his collection of diamonds was small—he found very few diamonds outside mathematics—and his claims for them modest—they were absolute truths, yes, but only about the world *as we experience it,* not the world as it *is.*

Yet in certain respects he was typical. While there was, as you might expect, considerable disagreement among the diamond-hunters about the diamonds themselves—different thinkers developed different tests for diamond-hood, to which others' collections were subjected; frequently zircons were found—in the core of *every* collection was geometry. Everyone agreed that in the *Elements*, at least, were genuine diamonds, and that any test which found the theorems of geometry to be zircons could not possibly be valid.

Consider Plato (427?–347 B.C.), the first diamond-hunter whose writings have survived more or less intact. Plato described geometry as "knowledge of what eternally exists" (*Republic* 527 B). (Plato's geometry text was, of course, not Euclid's, which had not yet been written, but one of the earlier *Elements*.) He found it perfectly apparent that the theorems of geometry were diamonds, and expected that his readers would, on reflection, agree. He devoted his attention instead to the problem of *how* such absolute truth is acquired.

Pondering the fact that no drawing of a circle was ever mathematically perfect, and so could not be the *real* circle studied by geometers, he was led to postulate a timelessly existing realm of archetypes, the World of Forms, in which he supposed our souls to have resided before we were born, and which he supposed to contain perfect examples of the objects that exist only imperfectly in the world we inhabit—including geometric objects like circles and straight lines. The postulates of geometry have the spontaneous (Kant would have said "a priori") appearance of truth, decided Plato, because they remind us of our prenatal existence in the World of Forms, during which we encountered real geometric objects behaving exactly as the postulates now say they should have. (Kant would have said that our memories of existence in the World of Forms was, for Plato, the *X* that accounted for the postulates' synthetic nature.)

Plato's philosophy is certainly very different from Kant's. Kant, with the experience of 2200 years of post-Platonic philosophers to draw on, gave up entirely the hope of knowing the world as it *is* and settled for a small collection of diamonds about the world we *experience.* Plato naively (Kant would have said) identified the two—what we experience *is* the world as it is—but, downgrading both, located complete reality only in the underlying World of Forms, descriptions of which constituted Plato's much larger collection of diamonds.

These differences notwithstanding, Plato and Kant, and most of the philosophers and scientists in the 2200-year interval between them, did share the following general presumptions:

(1) Diamonds—informative, certain truths about the world—exist.
(2) The theorems of Euclidean geometry are diamonds.

Presumption (1) is what I referred to earlier as the "Diamond Theory" of truth. It is far, far older than deductive geometry. Some scholars[5] argue there was a time, long before Thales, when people didn't realize that they themselves were the sources of the statements that came suddenly into their heads, so experienced their own thoughts as the voices of gods (and therefore as diamonds). There is wide agreement even among scholars who find this theory outlandish that before philosophy and science people experienced the world more *immediately* than we do today, and that consequently their myths, as expressions of this more direct experience, were reckoned as diamonds.

Deductive geometry was not, then, the origin of the Diamond Theory. It was, however, the principal factor by which the Diamond Theory was *maintained*, as an old habit, once the world no longer spoke directly and people had to turn to philosophy and science. From Plato to Kant, (2) was the strongest evidence for (1). Geometers had unearthed an imposing cache of what were, manifestly, diamonds. This was proof not only that diamonds existed, but also of the mind's power to discover them. There was no reason why investigation of other areas—cosmology, say, or ethics—could not achieve the same degree of success. Objective truth must lie all around, waiting to be discovered. The diamond-hunt was on.

As the diamond-hunters were the dominant shapers of the Western intellectual tradition, the impact of Euclidean geometry on Western culture can be seen to go far beyond that of a mathematical tool. Until the last century the Diamond Theory of truth pervaded not only philosophy and science, but also theology, history, politics, ethics, law, economics, aesthetics—indeed, *every* rational inquiry. It set the goals. Truths had to be independent of any human contribution, like diamonds dug out of the earth.

Notes

[1] *Immanuel Kant*. One of the greatest names in philosophy. Kant's biography is proof that you don't have to be flashy to become famous. Kant spent his entire life within a few miles of Königsberg, East Prussia (now Kaliningrad, USSR, on the Baltic, 40 miles north of the Polish border). He was a poor, frail, and temperamentally serious professor, inflexible in his daily habits (housewives set their clocks by his afternoon walks), who wrote his major works in a style described by one modern editor as "so lacking in imaginative warmth, lucidity and literary charm that the student is often tempted to abandon them in despair." That the student usually does not, however, is testimony to what he finds beneath that difficult surface: a broad-minded, honest and incisive thinker grappling with the main problems of human existence.

[2] "*a priori*" In English "a priori" (originally a Latin phrase) should be treated as one word. I pronounce it "AY pry OR eye."

[3] "*All bachelors are unmarried.*" The example is Stephen Barker's in *Philosophy of Mathematics* (Prentice-Hall, 1964), p. 7.

[4] "*diamonds*" The happy metaphor is Morris Kline's in *Mathematics in Western Culture* (Oxford, 1953), p. 430.

[5] *Some scholars.* E.g., Bruno Snell in *The Discovery of the Mind in Greek Philosophy and Literature* (1960; Dover reprint, 1982); Julian Jaynes in *The Origin of Consciousness in the Breakdown of the Bicameral Mind* (Houghton Mifflin, 1982).

CHAPTER 4

The Problem With Postulate 5

For 2100 years after the appearance of the *Elements*, a steady trickle of subtle thinkers were disturbed by Postulate 5. It wasn't as simple as the other axioms. No one doubted it was true, but it seemed out of place as a basic assumption.

At first the problem was perceived as simply aesthetic. Postulate 5 "sounded" more like a theorem than an axiom. The fact that its converse actually is a theorem (Theorem 17) reinforced the feeling that it need not be assumed, but could be proved. As mathematicians have always considered it "inelegant" to assume more than is absolutely necessary, this led to the conviction that Postulate 5 *should* be proved.

Later, because of a misinterpretation of Euclid's intentions, the problem acquired a philosophic dimension as well. Euclid himself seems *not* to have considered the truth of his postulates to be obvious. "As regards the postulates," concludes Sir Thomas L. Heath after a lengthy analysis (Heath's *Euclid*, pp. 117–124), "we may imagine [Euclid] saying:

Besides the common notions there are a few other things which I must assume without proof, but which differ from the common notions in that they are not self-evident. The learner may or may not be disposed to agree to them; but he must accept them at the outset on the superior authority of his teacher, and must be left to convince himself of their truth in the course of the investigation which follows."

But as Euclid never made this explicit, the idea inevitably arose that he had intended not only the common notions, but also the postulates, to be acceptable at the outset as obviously true—a standard, many would feel, the complicated Postulate 5 did not meet. This gave new urgency to the problem with Postulate 5: its status as an axiom was more than mathematically inelegant—it was philosophically objectionable.

A number of historians have inferred from the structure of Book I that Euclid himself was somehow uncomfortable with Postulate 5. The fact that until Theorem 29 no theorem depends on it, and thereafter *every* theorem except Theorem 31 depends on it, makes it appear as if he wanted to postpone using it as long as he could. Further, there are theorems that he takes the trouble to prove without Postulate 5, though he could have proven them much more easily had he waited for Postulate 5 to appear on the scene—like the AAS part of Theorem 26, which is an immediate consequence of Theorem 32.

Poseidonios

The first person we know definitely to have been uncomfortable with Postulate 5 was the philosopher, scientist, and historian Poseidonios[1] (c. 135–c. 51 B.C.), who suggested proving Postulate 5 by changing Definition 23 to read

Parallel straight lines are straight lines which, being in the same plane and being produced indefinitely in both directions, keep always the same distance between them.

The defining property is called "equidistance" and is certainly well in accord with our mental image of parallel lines. In fact I suspect a layman asked to explain "parallel" lines would sooner give Poseidonios's definition than Euclid's.

You may be puzzled when I speak of "proving Postulate 5." In a given system of geometry one of course does not prove a postulate. Poseidonios's plan was to *reorganize* Euclidean geometry by striking the statement known to us as "Postulate 5" from the list of axioms and placing it instead among the theorems where it seemed to belong. In Poseidonios's proposed system his "proof of Postulate 5" would be the proof of a theorem. When I refer to "proving Postulate 5" now or in the sequel, I mean "proving the statement that in Euclid's system was called 'Postulate 5'."

Poseidonios changed the definition of "parallel" because a proof of Postulate 5 using Euclid's definition somehow escaped him. How could this be? Don't the two definitions amount to the same thing?

A little reflection on this anomaly reveals that Poseidonios's plan, unless modified, will not work. He clearly *intended* his definition of "parallel" to "amount to the same thing" as Euclid's, that is, to apply in exactly the same situations. For example, if AB is a given straight line, he intended that every straight line "parallel" to AB in his sense of the word (equidistant from AB) would also be "parallel" to AB in Euclid's sense (never meeting AB), and vice-versa. Though it certainly seems impossible to imagine a straight line parallel to AB in one sense but not the other, the issue here is not what is imaginable, but what is logical.

To make things more explicit, Poseidonios's plan stands or falls according as the mathematical community is certain or doubtful of the following two statements:

(1) If two straight lines are equidistant no matter how much they are produced, then they never meet.

(2) If two straight lines never meet no matter how much they are produced, then they are equidistant.

Of course no one has ever doubted (1). If two straight lines keep always the same (positive) distance between them, then the distance between them can never be zero. But what of (2)? Straight lines can be produced indefinitely, so we can never examine them over their entire potential lengths. Even if we know a certain pair of straight lines never meet, no matter how much they are produced, and we find them to be equidistant over the tiny portion of their potential lengths to which we have access, how can we be sure the distance between them doesn't fluctuate somewhere beyond our reach? (If you answer, "Because they are *straight*!", remember that "straight line" is a primitive term and the only properties of primitive terms we can use are those contained in the axioms.) I am not quibbling. In mathematics a request for a supporting argument is *always* legitimate. We need a proof of (2).

We already have one, practically. That parallel lines (in Euclid's sense) are equidistant is an easy consequence of Theorem 34. But such a proof won't do here, as Theorem 34 depends on Postulate 5, the very postulate Poseidonios has struck from the roster!

Actually it is known that, given Poseidonios' foundation, *no* proof of (2) is possible. (2) must be *postulated*, as Poseidonios tacitly did when he assumed his definition of "parallel" to be equivalent to Euclid's. And since postulating (2) explicitly would make his new definition of "parallel" redundant, we may as well keep the old one.

The result, after the dust settles, is a plan for reorganizing Euclidean geometry that is practicable but less dramatic than Poseidonios' original proposal. In it Postulate 5 is still demoted to Theorem, but its place, which Poseidonios would have left vacant, has now been taken by statement (2), which from now on I will call "Poseidonios' Postulate."

Poseidonios' Reorganization (Modified)

foundation
{
Keep Euclid's primitive terms and Definitions of defined terms (including Euclid's Definition of "parallel");

keep Euclid's Common Notions, Postulates 1–4 and Postulates 6–10;

replace Postulate 5 with Poseidonios' Postulate, which (since we are keeping the original Definition of "parallel") now says: "Parallel straight lines are equidistant."
}

theorems
{
Keep the theorems proven without Postulate 5, with their original proofs;

state and prove a new theorem whose content is that of Euclid's Postulate 5;

keep the theorems Euclid proved *with* Postulate 5, with their original proofs, except that citations of "Postulate 5" are changed to citations of the new theorem.
}

Note that to effect the entire scheme it remains only to prove the "new theorem," i.e., to prove Postulate 5.

Proof of Postulate 5, After Poseidonios

None of Poseidonios' work is extant, and we know of his plan only through Proclus (p. 75). Further, we have modified his original proposal. What follows, then, is not his actual work, but merely my way of completing the reorganization outlined above. First an auxiliary theorem, which I'll call "Theorem P1" ("P" for "Poseidonios"), then the actual deduction of Postulate 5 ("Theorem P2"). In a Poseidonian geometry book these would come right before Theorem 29.

Theorem P1. *Through a given point not on a given straight line, and not on that straight line produced, no more than one parallel straight line can be drawn.*

Proof.

1. Let "*P*" be a point not on straight line "*AB*", and not on *AB* produced (see Figure 102). — hypothesis

2. Pretend that through *P* there are two straight lines "*Y*"*P*"*Z*" and "*W*"*P*"*X*", both parallel to *AB*. To be specific, say that to the left of *P*, *WPX* is the higher of the two. — RAA hypothesis

3. Choose "*C*" at random on *WP* and draw *C*"*E*" perpendicular to *AB* (or *AB* produced). Draw *P*"*Q*" perpendicular to *AB* (or *AB* produced). — Post 10, Th 12 (Post 2)

4. *CE* intersects *PY* (or *PY* produced) in a point "*D*". — 2, Post 7 (Post 2)

5. *CE* = *PQ* — Poseidonios' Post (*WP*∥*AB*)

6. *DE* = *PQ* — Poseidonios' Post (*YP*∥*AB*)

7. *CE* = *DE* — 5, 6, C.N. 1

8. But *CE* > *DE*. — C.N. 5

9. Contradiction. — 7 and 8

10. Therefore, through *P* no more than one parallel can be drawn. — 2–9, logic

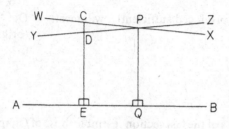

Figure 102

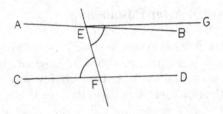

Figure 103

Theorem P2. *If a straight line falling on two straight lines make the interior angles on the same side less than two right angles, the two straight lines, if produced indefinitely, meet on that side on which are the angles less than the two right angles.*

Proof.

1. Let "*EF*" be a straight line falling on straight   hypothesis
   lines "*AB*" and "*CD*" so that (say) $\angle BEF +$
   $\angle EFD < 180°$ (see Figure 103).

(By Theorem 13 and the Common Notions, $\angle AEF + \angle EFC > 180°$, so Theorem 17 would be violated if *AB* and *CD* were to meet to the left of *EF*. Thus it suffices to show merely that *AB* and *CD* meet—that they do so to the right is immediate.)

2. Through *E* draw *E*"*G*" such that $\angle GEF =$ Th 23
   $\angle EFC$.

(Steps 3–5 show that *EG* is above *EB*, as drawn.)

3. $\angle EFC + \angle EFD = 180°$ Th 13

4. $\angle BEF < \angle EFC$ 1, 3, C.N. 6 (If $a + b$
                                                           $< c$ and $d + b = c$
                                                           then $a < d$.)

5. $\angle BEF < \angle GEF$ 2, 4, C.N. 6 (If $a = b$
                                                           and $c < b$ then $c < a$.)

(Thus *AB* and *EG* are distinct lines through *E*.)

6. $EG \| CD$ 2, Th 27

7. $AB \nparallel CD$ 5, 6, Th P1

8. *AB* and *CD*, if produced sufficiently, will meet.   Def "parallel"
                                                           (original Def 23)

Metageometry

Though the proofs of the last section resemble those of Chapter 2, our purpose in constructing them was different. We wanted to establish, not facts about straight lines in the plane, but rather that Euclid's Postulate 5 can be proven as a theorem in Poseidonios' reorganization—a fact about an *entire geometric system*. Modern mathematicians call such facts "*meta*geometrical."

In *geometry* we examine geometric figures and report (in "axioms" and "theorems") on what we see. Thus the objects of our study are *figures*, and the system we construct is a list of statements about those figures. In *metageometry*, on the other hand, the objects of study are *geometric systems* themselves, and so our reports on what we see ("metatheorems") are statements about *statements*. For example, "In isoceles triangles the angles at the base are equal" is a statement about geometric figures and is part of Euclid's system—it is, in fact, Theorem 5; but the statement, "Euclid's Theorem 5 can be proven without Postulate 5" is a metatheorem because it does not concern geometric figures (directly), but rather the logical relationship of certain statements *about* geometric figures in Euclid's system.

In this light the proofs of the last article can be seen as the proofs of two metatheorems:

Metatheorem 1. *In Poseidonios' reorganization it is possible to prove that*:

Through a given point not on a given straight line, and not on that straight line produced, no more than one parallel straight line can be drawn.

Metatheorem 2. *In Poseidonios' reorganization it is possible to prove Postulate 5.*

In metageometry the methods are familiar, but the perspective from which we work, and so the import of our work, are new. Renaming our results recognizes these differences.

Evaluation of Poseidonios' Reorganization

Poseidonios wanted to somehow "improve" Euclidean geometry by showing the complicated Postulate 5 to be a logical consequence of Euclid's other axioms. Feeling that his definition of "parallel" was equivalent to Euclid's, he thought he knew how to do this.

Unfortunately, as we have seen, his definition comprises not only Euclid's definition, but also a tacit assumption that Euclid's parallels are equidistant. Carrying out Poseidonios' reorganization, therefore, does not consist in deducing Postulate 5 from the other axioms alone, as he had thought, but in deducing it from the other axioms *plus* his unconscious assumption (Poseidonios' Postulate). This we have done; but how close have we come to achieving Poseidonios' goal? Is Poseidonios' organization of Euclidean geometry any "better" than Euclid's?

At first blush it seems that it is. Reading Poseidonios' Postulate and Euclid's Postulate 5 side by side, the former strikes us as simpler, easier to understand (it "sounds" more like an axiom), and something a novice would more quickly accept (it seems more "self-evident").

Recalling my earlier objection, however, we can see that Euclid's postulate, for all its complexity, may be the one that is more "evident." A straight line can be produced indefinitely, and we have access to only a limited portion of it. Given a pair of straight lines that do not meet, we could not possibly check that they are equidistant over their entire potential lengths, as Poseidonios' Postulate asserts, because we cannot travel (even via telescope) indefinitely far. Conceivably (though admittedly difficult to imagine) the distance between the lines might fluctuate somewhere beyond our reach. As this applies to *any* pair of straight lines that do not meet, we cannot directly verify a single instance of Poseidonios' Postulate. We can, however, verify *some* instances of Euclid's postulate. Given two straight lines cut by a third so that the interior angles on the same side add up to less than 180°, Euclid's postulate asserts that the two straight lines will meet at a point that is only a finite distance away; if the angle-sum is not too close to 180° we will be able to travel to that point and see for ourselves.[2] And having verified the postulate whenever the point of intersection is reachable, we will tend to find it plausible even when the angle-sum is almost 180° and the point impracticably far.

Up to now we have been comparing the two postulates in psychological terms. But on mathematical issues the deciding factor is usually logical, so let's see what logicians have to say.

Modern logicians call two statements arising in some context "logically equivalent" if, in that context, each is deducible from the other. That is, if the two statements are A and B, then A and B are *logically equivalent* if

(1) Context + $A \Rightarrow B$, and
(2) context + $B \Rightarrow A$.

(The meaning of the double-shafted arrow—read "implies"—is that the statement it points to can be deduced from those before it.) In this situation logicians use the word "equivalent" (from Latin, "of equal value") because whenever A is true, so is B—by (1)—and whenever B is true, so is A—by (2)—so in the given context they are *equally* true; the truth of one is *related* to the truth of the other, and logicians don't care about anything (professionally!) except how statements are *related*. It doesn't matter if A and B seem to say very different things. Logic is blind to a statement's apparent substance, or tone, or complexity, or intuitive impact, or any other psychological feature. In deduction the only significant question one can ask about a statement is, "In view of what we have, does it follow?" A and B have the same logical value because, in the given context, the answer to that question for A would be the same as the answer for B.

Presently the context is the common part of Poseidonios' and Euclid's foundations—Euclid's primitive terms, Definitions of defined terms, Common Notions, and Postulates other than #5, to which we can add all the theorems that are deducible therefrom. I will call this context "Neutral geometry,[3]" as it is the part of Euclidean geometry that is noncontroversial. The statements whose relative logical value we would like to assess are Poseidonios' Postulate and Euclid's Postulate 5. They will be logically equivalent if

(1) Neutral geometry + Poseidonios' Postulate $\Rightarrow$ Euclid's Postulate 5, and
(2) Neutral geometry + Euclid's Postulate 5 $\Rightarrow$ Poseidonios' Postulate.

We already have implication (1), because what it asserts is precisely what occupied us two sections ago ((1) is Theorem P2). We practically have implication (2) as well. We said on p. 120 that the equidistance of parallel straight lines in Euclid's system is an easy consequence of Theorem 34, so verifying implication (2) is just a matter of writing down those few steps, which I leave to you. Thus we have

Metatheorem 3. *In the context of Neutral geometry, Poseidonios' Postulate and Euclid's Postulate 5 are logically equivalent.*

Had implication (2) turned out to be false, Poseidonios' Postulate would have been logically prior, more "basic" (and more "sweeping") than Euclid's. It would have entailed Euclid's but Euclid's would not have entailed it. In that case substituting Poseidonios' Postulate for Euclid's would have had considerable value by simplifying the logical structure of Euclidean geometry's foundation, but as things have turned out, a logician would say there is simply no point to adopting Poseidonios' reorganization.

We have considered three viewpoints, each coming to its own conclusion on the relative merit of Poseidonios' and Euclid's systems. To summarize: the first view—which without prejudice I will call the "naive" view—is that Poseidonios' Postulate is better because it is shorter and easier to comprehend (more as an axiom "should" be); the second—I will call it the "scientific" view—is that Euclid's postulate is better because it is experimentally verifiable at least in part; and the third or "logical" view is that neither is preferable because they are logically equivalent.

The choice is not an easy one. Behind the various views are different opinions—all well-established—as to the very nature of Euclid's work. Is the *Elements* (a) a textbook, or (b) a work of art? Is it (c) a scientific description of space? Or (d) an exercise in pure logic? From the beginning all four aspects have been present, with different emphases, in the ways scholars have perceived the book.

It's easy to see how believing the *Elements* to be primarily (a) or (b) leads to the naive viewpoint. For then the difficulty with Postulate 5 is either pedagogical (it is difficult for a beginner) or aesthetic (it is a blemish on Euclid's beautiful system), but in any case bound up with Postulate 5's *unwieldiness*. Poseidonios' Postulate is then an improvement. Similarly finding the *Elements* to be primarily (c) causes one to interpret the controversy as concerning the *reliability of a scientific principle* and so tilts the balance in favor of a postulate subject to experimental test. This is the scientific view, and favors

Euclid's postulate. Finally, believing the *Elements* to be primarily (d) translates the discomfort with Postulate 5 into a longing for a *logically simpler foundation* for geometry. This is the logical view, according to which replacing Postulate 5 with a logically equivalent statement (such as Poseidonios' Postulate) is pointless.

Let me caution you that I'm making the alternatives seem fewer and more clear-cut than they really are. The naive, scientific and logical views and the corresponding perceptions of the *Elements* are only *general postures* I've proposed to call attention to the various issues that arise when we contemplate substituting another postulate (and not necessarily Poseidonios') for Postulate 5. The actual viewpoints of living, thinking people can't always be placed neatly in categories. A person might, for example, consider the *Elements* to be primarily a treatise on physical space ("scientific view"), but still favor Poseidonios' Postulate on grounds that in my categorization are associated with the other perspectives. His/her reasoning might go like this:

A. Principles should be as simple as possible ("naive view").
B. Since Poseidonios' Postulate is logically equivalent to Euclid's postulate ("logical view"), an experiment tending to confirm one simultaneously tends to confirm the other.
C. Thus the experimental advantage of Euclid's postulate is illusory.
D. Therefore, other things being equal, Poseidonios' Postulate is better because it is simpler.

And so on. You may yourself have forged a chain of reasoning, in support of one postulate or the other, that is different still.

The only way to settle the question is by fiat. I will, from now on in this chapter, concern myself chiefly with the posture I have called the "logical view." I will do this not because I know of any crippling objections that can be made to the other viewpoints, nor because I have been holding in reserve some overwhelming argument in its favor, but simply because it is the viewpoint that *kept the controversy alive* and thereby led, ultimately, to non-Euclidean geometry.

Overview of Later Attempts

Between the first century B.C. and the early 19th century scores of thinkers were disturbed by Postulate 5 and set about trying to prove it. Those who succeeded invariably made use of an extra assumption. Consequently all the many proposals boil down to the same general plan we uncovered in our analysis of Poseidonios' proposal:

(1) Replace Postulate 5 with a more acceptable assumption;
(2) leave the rest of Euclid's foundation (i.e., the foundation of Neutral geometry) intact;
(3) prove Postulate 5.

Often, especially at first, the replacement postulate was substituted unconsciously, and for a while people thought (as had happened to Poseidonios) that Postulate 5 had been deduced from the Neutral foundation *alone*. This would have meant that the troublesome Postulate 5 could simply be removed from Euclidean geometry's foundation, that what remained would be able to support the entire superstructure of theorems unassisted, and thus that Euclidean geometry and Neutral geometry were the same. But in every case the hidden assumption was eventually ferreted out by some later thinker.

Other times the mathematician was well aware of his replacement postulate, but felt he had found one with which the mathematical world could at last be content. The English mathematician John Wallis (1616–1703), for example, proposed replacing Postulate 5 with the following.

Wallis' Postulate. On a given finite straight line it is always possible to construct a triangle similar to a given triangle.[4]

("Similar" triangles have the same three angles, and so the same shape.) You may find it difficult to imagine how Wallis could have deduced Postulate 5 from this, as it seems to say a very different thing; but logically it is not much further from Postulate 5 than Poseidonios' Postulate was. Wallis defended his postulate on the ground that what it asserts for triangles is very nearly what Postulate 3 already does for circles, i.e., the existence of figures having the same shape but arbitrary size.

A few mathematicians, especially during the 18th and early 19th centuries, were less easily satisfied.

As for me, I have already made some progress in my work. However the path I have chosen does not lead at all to the goal which we seek [proving Postulate 5] . . . . It seems rather to compel me to doubt the truth of geometry itself.

It is true that I have come upon much which by most people would be held to constitute a proof: but in my eyes it proves as good as *nothing*. For example, if one could show that a . . . triangle is possible, whose area would be greater than any given area, then I would be ready to prove the whole of geometry absolutely rigorously.

Most people would certainly let this stand as an Axiom; but I, no! It would, indeed, be possible that the area might always remain below a certain limit, however far apart the three angular points of the triangle were taken.

—letter[5] from Karl Friedrich Gauss
to Farkas Bolyai, December 17, 1799

(People who like to compare mathematicians frequently rank the German mathematician, physicist, and astronomer Gauss (1777–1855) with Archimedes and Isaac Newton in a three-way tie for all-time greatest. You will see Gauss' name again: a few years after writing this letter, he became one of the inventors of non-Euclidean geometry. Gauss' lifelong friend Farkas Bolyai (1775–1856) was a Hungarian mathematician whose son, János Bolyai (1802–1860), became another of the inventors of non-Euclidean geometry.) Gauss' replacement postulate I will naturally call

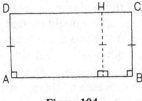

Figure 104

Gauss' Postulate. It is possible to construct a triangle whose area is greater than any given area.

Psychologically this is even further from Postulate 5 than Wallis' Postulate was, and now the logical distance is great as well; nonetheless Gauss did manage, amazingly, to deduce Postulate 5. But note his displeasure at having to resort to *any* replacement postulate *at all*.

Here are some of the more noteworthy replacement postulates that have been proposed over the years, grouped by surface similarity; most are followed by the names of the mathematicians who proposed them.

1A. Parallel straight lines are equidistant. (Poseidonios, 1st century B.C.)

1B. All the points equidistant from a given straight line, on a given side of it, constitute a straight line. (Christoph Clavius, 1574)

1C. In any quadrilateral having two equal sides perpendicular to a third side (Figure 104), there is at least one point (*H* in the figure) on the fourth side from which the perpendicular to the side opposite is equal to the two equal sides. (Giordano Vitale, 1680)

1D. There exists at least one pair of equidistant straight lines.

2. The distance between a pair of parallel infinite straight lines (may fluctuate, but) remains less than a certain fixed distance. (Proclus, 5th century)

3A. (Theorem 30) Straight lines parallel to the same straight line are also parallel to one another.

3B. If a straight line intersects one of two parallel straight lines, it will, if produced sufficiently, intersect the other (or the other produced) also.

3C. Through a given point not on a given straight line, and not on that straight line produced, no more than one parallel straight line can be drawn. (popularized by John Playfair, late 18th century)

4A. If two straight lines (*AB*, *CD* in Figure 105) are cut by a third (*PQ*) that is perpendicular to only one of them (*CD*), then the perpendiculars from *AB* to *CD* are less than *PQ* on the side on which *AB* makes an acute angle with *PQ*, and are greater on the side on which *AB* makes an obtuse angle. (Nasir al-Din, 13th century)

4B. Straight lines which are not equidistant converge in one direction and diverge in the other. (Pietro Antonio Cataldi, 1603)

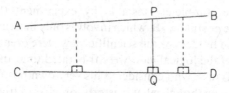

Figure 105

5A. On a given finite straight line it is always possible to construct a triangle similar to a given triangle. (John Wallis, 1663; Lazare-Nicholas-Marguerite Carnot, 1803; Adrien-Marie Legendre, 1824)

5B. A pair of noncongruent, similar triangles exists. (Gerolamo Saccheri, 1733)

6A. In any quadrilateral having two equal sides perpendicular to a third side (Figure 104), the other two angles are right. (Gerolamo Saccheri, 1733)

6B. In any quadrilateral having three right angles, the fourth angle is also right. (Alexis-Claude Clairaut, 1741; Johann Heinrich Lambert, 1766)

6C. At least one rectangle exists. (Gerolamo Saccheri, 1733)

7A. (Theorem 32 (b)) The angle-sum of every triangle is 180°. (Gerolamo Saccheri, 1733; Adrien-Marie Legendre, early 19th century)

7B. At least one triangle with angle-sum 180° exists. (same)

8. There is no absolute standard of length. (Johann Heinrich Lambert, 1766; Adrien-Marie Legendre, early 19th century)

9A. Every straight line through a point within an angle will, if produced sufficiently, meet at least one side of the angle (or one side produced). (J. F. Lorenz, 1791)

9B. Through any point within an angle of less than 60°, it is always possible to draw a straight line which meets both sides (produced if necessary) of the angle. (Adrien-Marie Legendre, early 19th century)

10. It is possible to construct a triangle whose area is greater than any given area. (Karl Friedrich Gauss, 1799)

11. In the plane translations and rotations of a straight line are independent motions. (Bernhard Friedrich Thibaut, 1809)

12. Given three points not on a single straight line, it is always possible to draw a circle passing through all three. (Adrien-Marie Legendre, Farkas Bolyai, early 19th century)

Over the years holders of the naive view, for whom the chief difficulty with Postulate 5 was its unwieldiness, were relatively quickly satisfied, if not by Poseidonios' Postulate—which they may not have heard of—then by some other concise alternative (for instance "Playfair's Postulate" 3C, which has been widely used in textbooks). And substitutes for Postulate 5 were discovered that were completely testable by experiment (for instance the Saccheri–Legendre postulate 7B, which involves only measuring the angles of a *single* triangle), so holders of the scientific view were eventually satisfied as well. But holders of the logical view were frustrated time and again.

Let's return to Gauss, who considered his very own postulate "as good as nothing" (p. 127), even though what it asserts for areas of triangles—that they can be made arbitrarily large—is no more than what Postulate 2 asserts for lengths.

Gauss realized, I think, that his postulate is logically equivalent to Postulate 5. Remember that a statement "A" is logically equivalent to Postulate 5 if

(1) Neutral geometry + $A \Rightarrow$ Postulate 5, and
(2) Neutral geometry + Postulate $5 \Rightarrow A$,

where "Neutral geometry" consists of Euclid's primitive terms, Definitions of defined terms, Common Notions, Postulates other than #5, and all theorems derivable therefrom. Taking "A" to be Gauss' Postulate, Gauss himself had established (1) when he proved Postulate 5. And, although we didn't go that far in Chapter 2, it is well-known that in Euclid's geometry one can construct a triangle with area as large as one pleases, which (since Neutral geometry + Postulate 5 = Euclid's geometry) is what (2) says.

You may wonder why Gauss, if he realized the logical equivalence, didn't say so in his letter to Farkas Bolyai. He didn't because he didn't have the vocabulary. In mathematics it takes a long time for a new abstract concept to emerge and then come into focus, and until this happens it can't be sharply defined. "Logical equivalence of geometric postulates" is a notion not formulated precisely until long after Gauss wrote his letter, at a time when mathematicians were shocked by non-Euclidean geometry into looking at all geometry more abstractly than ever before. When I attribute the "logical view" to Gauss and other mathematicians of the early 19th century and before (especially Saccheri, Lambert, Legendre, and Farkas Bolyai), I am interpreting their work in modern terms they would never have used. Nonetheless their espousal of the logical view is apparent in their continual dissatisfaction with replacement postulates, and their dogged striving to prove Postulate 5 without resorting to one.

This persistent unwillingness of Gauss and his intellectual brothers to accept a replacement postulate indicates that they were aware, though we can only guess how clearly, of a surprising thing I haven't mentioned yet, although it may have already occurred to you: that *any* replacement postulate (any of the ones listed on pp. 128–129, for example) is logically equivalent to Postulate 5! Nowadays this is easy to see, in fact by practically the same argument we used two paragraphs ago. For whoever proposes a replacement postulate "A" will of course use it to prove Postulate 5—that's the whole point—and thereby establish (1). But the proposed postulate A will surely be something that is provable in Euclidean geometry (as originally organized), and thus (2) will occur as well! Thus *no* replacement postulate can be logically prior to

Postulate 5; they are *all* logically equivalent to it and therefore, in the logical view, *all* worthless.

Suddenly the frustration of Gauss and the other logical-view mathematicians is understandable. They could see no way of proving Postulate 5 without a replacement postulate. Yet—and in varying degrees they sensed this—no replacement postulate, even one of immense intuitive appeal, could possibly serve, for no matter how preferable it was psychologically, it would be equivalent to Postulate 5 and therefore indistinguishable from it where it counted—logically.

So Near

The main thread of my story continues in the next chapter. If you are impatient to get on with it, or just not in the mood for a technical discussion that is only tangential, you may want to skip this and the remaining section of this chapter. Eventually you will have to read small parts of the present section, as our development of non-Euclidean geometry will depend on them; but they can be postponed, and when the time comes I'll refer you back.

In this section I will establish connections among some of the replacement postulates listed on pages 128–129, in order to show you how extremely near to success mathematicians had come by 1800 in their campaign to prove Postulate 5 without an extra assumption, and incidentally to give you more experience with the strange self-denial trying to avoid an extra assumption involves.

In the process we will also derive a completely practical experimental test of Postulate 5, which I will discuss in the following section. You may want to read that section (it begins on page 147) even if you decide to skip this one.

Much of what follows is taken, in outline, from a book called *Euclides ab Omni Naevo Vindicatus* ("Euclid Freed of Every Flaw," Milan, 1733; an English translation by George B. Halsted was reprinted by Chelsea, 1986) by the University of Pavia mathematics professor Gerolamo Saccheri (1667–1733), the most sustained attempt at proving Postulate 5 to appear up to that time.

With Saccheri we take as our starting point *only* Neutral geometry: Euclid's primitive terms, Definitions of defined terms, Common Notions, Postulates other than #5, and theorems proven without Postulate 5. We will endeavor to make no other assumption. This will at times be difficult; overcoming an unconscious habit usually is. (I was once a nail biter, and remember how insidiously the habit reasserted itself whenever my vigilance slackened, as when reading or watching TV.) The problem is simply that the Postulate 5-dependent theorems assert things that we judge to be *true*—raised on Euclidean geometry, it is difficult for us to imagine any other state of affairs—and we will instinctively tend to inject these beliefs into our investigation.

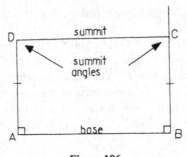

Figure 106

Pursuing the project faithfully will require that we take the extreme measure of *shutting out* the entreaties of our intuitions and imaginations—a forced separation of mental powers that will quite understandably be confusing and difficult to maintain, as it amounts to simulated madness.

Central to Saccheri's work is a peculiar quadrilateral now named for him.

Definition. A *Saccheri quadrilateral* is a quadrilateral in which a pair of opposite sides are equal and have one of the other sides as a common perpendicular (see Figure 106). The common perpendicular is called the *base*, the side opposite it is the *summit*, and the angles adjacent to the summit are the *summit angles*.

The common perpendicular is always called the "base," even if in a particular drawing it is elsewhere than at the bottom; similarly for the "summit."

You should resist the temptation to think of a Saccheri quadrilateral as a rectangle.

Definition. A *rectangle* is a quadrilateral with four right angles.

While it's true that in Euclidean geometry all Saccheri quadrilaterals are rectangles and vice-versa (which is why you've never heard the term "Saccheri quadrilateral"), the proof that this is so is unusable in the present context because it depends on Postulate 5. In fact, since the only construction of a rectangle we have seen (Theorem 46) also uses Postulate 5, in Neutral geometry the very existence of rectangles is thus far an open question.

On the other hand there's no question that in Neutral geometry Saccheri quadrilaterals exist (see Figure 106). Choose points *A* and *B* at random in the plane (Postulate 10); draw *AB* (Postulate 1); draw *AD* perpendicular to *AB* (Theorem 11); draw another perpendicular from *B* (Theorem 11), produce it if necessary (Postulate 2), and cut off a length *CB* equal to *DA* (Theorem 3); draw *DC* (Postulate 1), and *ABCD* is a Saccheri quadrilateral with base *AB*, summit *DC*, and summit angles *CDA* and *DCB*.

Most things that can be established about Saccheri quadrilaterals in Neu-

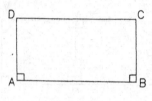

Figure 107

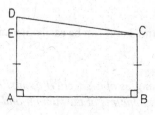

Figure 108

tral geometry are consequences of the following rather general theorem, which we will make further use of later on.

Theorem A. *If the vertices of a quadrilateral are consecutively A, B, C, D with right angles at A and B* (Figure 107), *then DA is greater than, equal to, or less than CB according as, respectively, ∠CDA is less than, equal to, or greater than ∠DCB.*

Proof. There are six things to prove.

Case 1. DA = CB. (To show: ∠CDA = ∠DCB.) I'll leave this to you as an exercise. It's easy to do with triangles. One thing, though: you'll need to use more than two. If you think you've done it with just two triangles, you've assumed something not yet established as part of Neutral geometry.

Case 2. DA > CB (see Figure 108). (To show: ∠CDA < ∠DCB.)

1. Pick "E" on DA so that EA = CB.	Th 3
2. Draw EC.	Post 1
3. ∠CDA < ∠CEA	Th 16
4. ∠CEA = ∠ECB	Case 1 (applied to ABCE)
5. ∠ECB < ∠DCB	C.N. 5
6. ∠CDA < ∠DCB	3–5, C.N. 6 (If a < b, b = c, c < d then a < d.)

Case 3 (given *DA < CB*, show ∠CDA > ∠DCB) is similar. Case 4 (given ∠CDA = ∠DCB, show *DA = CB*), case 5 (given ∠CDA < ∠DCB, show *DA > CB*), and case 6 (given ∠CDA > ∠DCB, show *DA < CB*) are the

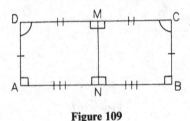

Figure 109

converses of cases 1, 2, and 3 and can be easily proven by contradiction using cases 1, 2, and 3.

Corollary. *The summit angles of a Saccheri quadrilateral are equal, and the straight line joining the midpoints of the summit and base is perpendicular to both.*

Proof. The first part is simply case 1; the second is an easy exercise.

Figure 109 summarizes what has been established about Saccheri quadrilaterals so far. Note that we have *not* proven a number of things that our Euclidean intuitions urge upon us: that the summit is equal to the base, that the summit angles are right, or that $MN = CB$. (On these points I have marked the figure in a noncommittal way.) It so happens that proving *any one* of these things would lead to a proof of Postulate 5! Saccheri knew this, and concentrated on trying to prove that the summit angles are right. To this end he made considerable progress, but before discussing his accomplishment let's show that success would indeed enable us to prove Postulate 5.

We want to show, then, that in the context of Neutral geometry, the hypothesis "the summit angles of a Saccheri quadrilateral are right" (which is replacement postulate 6A on p. 129) implies Postulate 5. This amounts to showing that 6A is logically equivalent to Postulate 5, as the reverse implication—Neutral geometry + Postulate 5 ⇒ 6A—is a matter of only a few steps. In the process we will also show that replacement postulate 7A (p. 129) is logically equivalent to Postulate 5.

Like Theorem A, the next theorem will also be used several times in the future.

Theorem B. *If, from the endpoints of a given side of a triangle (in Figure 110 the triangle is ABC and the given side is BC), perpendiculars (BF and CG) are drawn to a straight line (l) through the midpoints (D and E) of the other two sides, forming a quadrilateral (GFBC), then*

(1) *the quadrilateral is a Saccheri quadrilateral whose summit is the given side of the triangle (BC);*

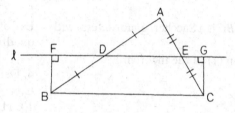

Figure 110

(2) *its base (FG) is twice the length of the straight line (DE) joining the midpoints of the triangle's other two sides; and*

(3) *its two summit angles (∠ FBC and ∠ GCB) have the same sum as the three interior angles of the triangle.*

Proof.

1. Let "*ABC*" be a triangle in which two sides— say *AB* and *AC*—have been bisected, respectively, at "*D*" and "*E*", *DE* has been drawn and produced in both directions, and straight lines *B*"*F*" and *C*"*G*" have been drawn perpendicular to this straight line.   hypothesis

2. Draw *A*"*H*" perpendicular to the straight line through *D* and *E*.   Th 12

(There are three cases to consider, according to whether *AH* is within, along one side of, or outside △*ABC*; according to, in other words, where the point *H* is situated relative to *D* and *E*.)

Case 1. *H* is between *D* and *E*.

3. Then *F* is on the other side of *D* from *H*, and *G* is on the other side of *E* from *H*, as shown in Figure 111.   exercise

4. Triangles *BFD* and *AHD* are congruent.   1, Post 4, Th 15, AAS
5. *FB* = *AH*   Def "congruent"
6. Triangles *CGE* and *AHE* are congruent.   1, Post 4, Th 15, AAS
7. *GC* = *AH*   6, Def "congruent"

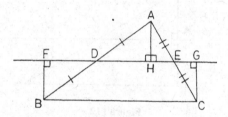

Figure 111

8. $FB = GC$ — 5, 7, C.N. 1

9. Therefore $GFBC$ is a Saccheri quadrilateral with summit BC. — Def of "Saccheri quadrilateral"

(This is conclusion (1) of the theorem.)

10. $FD = DH$ — 4, Def "congruent"

11. $DH = DH$ — C.N. 4

12. $FH = 2 \cdot DH$ — 10, 11, C.N. 2

13. Similarly, $HG = 2 \cdot HE$. — Imitate 10–12

14. Therefore $FG = 2 \cdot DE$. — 12, 13, C.N. 2

(This is conclusion (2).)

15. $\angle DBF = \angle DAH$ — 4, Def "congruent"

16. $\angle DBF + \angle ABC = \angle DAH + \angle ABC$ — 15, C.N. 6 (If $a = b$ then $a + c = b + c$.)

17. $\angle ECG = \angle EAH$ — 6, Def "congruent"

18. $\angle ECG + \angle ACB = \angle EAH + \angle ACB$ — 17, C.N. 6 (same as 16)

19. Therefore $\angle FBC + \angle GCB = \angle BAC + \angle ABC + \angle ACB$ — 16, 18, C.N. 2

(This is conclusion (3).)

Case 2. H coincides with D or E. Say H coincides with D.

20. Then F also coincides with D, and G is on the other side of E from D, as shown in Figure 112.

(Finishing Case 2 is an exercise. Be careful not to assume $\angle B$ is right.)

Case 3. H is on ED produced or DE produced. Say H is on ED produced.

21. Then F is on the other side of D from H, and G is on the other side of E from D, as shown in Figure 113. — exercise

(The figure shows F to be in fact between D and E, but all that can be proved is that it is somewhere between D and G. Fortunately, in proving Case 3 it makes no difference whether F coincides with E or is between E and G. As the proof follows the same general lines as the proof of Case 1, I'll leave the details for you.)

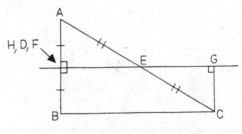

Figure 112

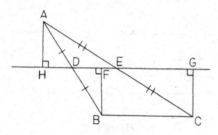

Figure 113

Metatheorem 4. *If in the context of Neutral geometry it were known that the summit angles of every Saccheri quadrilateral are right, then it would follow that the angle-sum of every triangle is 180°.*
(That is, Neutral geometry + replacement postulate 6A ⇒ replacement postulate 7A.)

Proof.
1. The summit angles of every Saccheri quadri-   hypothesis
   lateral are right.
2. Let "*ABC*" be any triangle.   hypothesis of repl.
                                                        post 7A
(To show: the angle-sum of △*ABC* = 180°.)
3. Bisect *AB* at "*D*" and *AC* at "*E*".   Th 10
4. Draw *DE* and produce it in both directions.   Post 1, Post 2
5. Draw *B"F"* and *C"G"* perpendicular to *DE*.   Th 12
(The result will look like Figure 111, Figure 112 or Figure 113, without *AH*.
In any case,)
6. *GFBC* is a Saccheri quadrilateral with summit *BC* and ∠*FBC* + ∠*GCB* = angle-sum of
   △*ABC*.
7. But ∠*FBC* + ∠*GCB* = 180°,   1, C.N. 2
8. so the angle-sum of △*ABC* = 180°.   6, 7, C.N. 1

Metatheorem 5. *If in the context of Neutral geometry it were known that the angle-sum of every triangle is 180°, then it would follow that in every triangle every exterior angle is equal to the sum of the two interior and opposite angles.*
(That is, Neutral geometry + 7A ⇒ Th 32(a).)

The proof is an exercise.

Geometric magnitudes obey the same laws as positive numbers. This is Common Notion 6, and so far every numerical law we've cited has been simple. The following[6] is less so:

If $a > b$, then there exists a positive whole number n such that $a/2^n < b$.

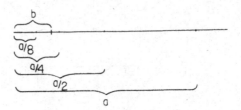

Figure 114

As we will need this to prove Metatheorem 6, I thought I would pause and discuss it first. (The proof of Metatheorem 6 is the only time we will use a complicated numerical law.) "$a/2^n$" signifies the result of successively halving n times, starting with a. Halving a gives $a/2$; halving $a/2$ gives $a/4 = a/2^2$; halving a third time gives $a/8 = a/2^3$, and so on. What the law asserts, therefore, is simply that by successively halving the greater (a) of two unequal positive quantities sufficiently often, one will eventually obtain a quantity smaller than the lesser (b). In Figure 114, a and b are lengths and $n = 3$. In the proof of Metatheorem 6, a and b will be angle sizes.

Note that the conclusion of the law would be true even if the hypothesis ($a > b$) were not. If a were $\leq b$, $a/2^n < b$ would be true for *every* positive whole number n, in particular for $n = 1$. Thus we could generalize our law into

For any a and b, there exists a positive whole number n such that $a/2^n < b$.

It is this generalization we will actually use in our proof.

Metatheorem 6. *If in the context of Neutral geometry it were known that*

in every triangle every exterior angle is equal to the sum of the two interior and opposite angles,

then it would follow that

through a given point not on a given straight line, and not on that straight line produced, it is always possible to draw a straight line making with the given straight line an angle smaller than any given angle.

(Let's call the part beginning with "through," the "Smaller Angle Property," S.A.P. for short. Then the metatheorem says that Neutral geometry + Th 32(a) $\Rightarrow$ S.A.P.).

Proof.
1. In every triangle every exterior angle is equal to   hypothesis
   the sum of the two interior and opposite angles.
2. Let "P" be a given point not on a given straight   hypothesis of S.A.P.
   line "l", and not on l produced, and let "ABC"
   be a given angle (see Figure 115).

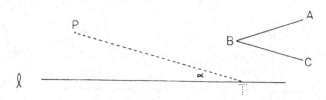

Figure 115

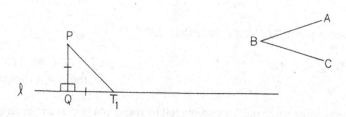

Figure 116

(To show: through P it is possible to draw a straight line $P“T”$ making an angle "α" $< \angle ABC$.)

3. There is a positive whole number n such that $90°/2^n < \angle ABC$.

    C.N. 6 (For any a and b, there exists a positive whole number n such that $a/2^n < b$.)

4. Draw $P“Q”$ perpendicular to l (produced if necessary).

    Th 12 (Post 2)

5. From l, or l produced, cut off $Q“T_1” = PQ$ (see Figure 116).

    Th 3 (Post 2)

6. Draw PT_1.

    Post 1

7. $\angle T_1PQ = \angle PT_1Q$

    Th 5

8. $\angle T_1PQ + \angle PT_1Q = 90°$

    1

9. $\angle PT_1Q = 90°/2$

    7, 8, C.N. 6 (If $a = b$ and $a + b = c$ then $b = \frac{1}{2}c$.)

10. From QT_1 produced cut off $T_1T_2 = PT_1$, and draw PT_2 (see Figure 117).

    Th 3 (Post 2), Post 1

11. $\angle T_2PT_1 = \angle PT_2Q$

    Th 5

12. $\angle T_2PT_1 + \angle PT_2Q = 90°/2$

    1, 9

13. $\angle PT_2Q = 90°/2^2$

    11, 12, C.N. 6 (If $a = b$ and $a + b = c$ then $b = \frac{1}{2}c$.)

14. Similarly, by repeating the sequence we have already gone through twice, we will eventually

    5–9, 10–13

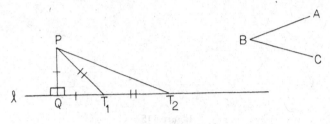

Figure 117

locate a point "T_n" on l such that $\angle PT_nQ =$
$90°/2^n$.

15. $\angle PT_nQ < \angle ABC$ 3, 14, C.N. 6 (If $a < b$
                                                      and $c = a$ then $c < b$.)

The preceding three metatheorems tell us respectively that in the context of Neutral geometry

(1) replacement postulate 6A $\Rightarrow$ replacement postulate 7A,
(2) replacement postulate 7A $\Rightarrow$ Theorem 32(a), and
(3) Theorem 32(a) $\Rightarrow$ the Smaller Angle Property.

Taking them all together we see that

Corollary. *Neutral geometry* $+ 6A \Rightarrow 7A$, *Th 32(a), and* S.A.P.

We are now ready to prove the result we have been steering for.

Metatheorem 7. *If in the context of Neutral geometry it were known that the summit angles of every Saccheri quadrilateral are right, then it would be possible to prove Postulate 5.*
(That is, Neutral geometry $+ 6A \Rightarrow$ Postulate 5.)

Proof.
1. The summit angles of every Saccheri quadri-   hypothesis
   lateral are right.
2. Let "EF" be a straight line falling on straight   hypothesis of Post 5
   lines "AB" and "CD" such that $\angle BEF +$
   $\angle EFD < 180°$ (see Figure 118).
(To show: AB and CD, if produced sufficiently, will meet to the right of EF.)
3. Through E draw $E"G"$ making $\angle GEF = \angle CFE$.   Th 23
4. $\angle GEF > \angle BEF$, as drawn.   exercise
5. Draw $E"H"$ from E to CD (produced if neces-   above Cor (S.A.P.
   sary) so that $\angle EHF < \angle GEB$.   with E the given
                                                          point, CD the given
                                                          straight line, and
                                                          $\angle GEB$ the given
                                                          angle)

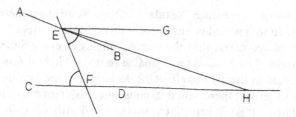

Figure 118

(To show: *AB* enters ∠*FEH*, as drawn. This can be accomplished by showing ∠*GEH* < ∠*GEB*. It will then follow by Postulate 2, Postulate 6(iii) and Postulate 1—see the discussion after the proof of Theorem 10, p. 65—that *AB*, if produced sufficiently, will intersect *CD*, or *CD* produced, between *F* and *H* and we will be done.)

6. ∠*CFE* = ∠*FEH* + ∠*EHF*	1, above Cor (Th 32 (a))
7. ∠*GEF* = ∠*FEH* + ∠*GEH*	C.N. 4
8. ∠*CFE* = ∠*FEH* + ∠*GEH*	3, 7, C.N. 1
9. ∠*EHF* = ∠*GEH*	6, 8, C.N. 6 (If *a* = *b* + *c* and *a* = *b* + *d* then *c* = *d*.)
10. ∠*GEH* < ∠*GEB*	5, 9, C.N. 6 (If *a* < *b* and *a* = *c* then *c* < *b*.)
11. Therefore *AB*, if produced sufficiently, will intersect *CD*, or *CD* produced, between *F* and *H*.	Post 2, Post 6, Post 1

Corollary 1. *In the context of Neutral geometry the statement that the summit angles of every Saccheri quadrilateral are right (6A) is logically equivalent to Postulate 5.*

Proof. The claim is that

Neutral geometry + 6A ⇒ Post 5, and
Neutral geometry + Post 5 ⇒ 6A.

The metatheorem gives us the first implication. Verifying the second is an exercise.

Corollary 2. *In the context of Neutral geometry the statement that the angle-sum of every triangle is 180° (7A) is logically equivalent to Postulate 5.*

Proof. That Neutral geometry + 7A ⇒ Post 5 follows from Metatheorems 5 and 6 and steps 2–11 of the proof of Metatheorem 7 (which show that Neutral geometry + Th 32(a) + S.A.P. ⇒ Post 5). Our proof of Theorem 32 (b) (p. 89) establishes that Neutral geometry + Post 5 ⇒ 7A.

Saccheri wanted to deduce Postulate 5 from Neutral geometry *alone*. He knew that to do so it would be sufficient—this is Metatheorem 7—to deduce, within Neutral geometry, that the summit angles of every Saccheri quadrilateral are right. This Saccheri was unable to do, though at one point in his book, exhausted by prodigious effort, he declares that the summit angles *must* be right, for otherwise there would be consequences "repugnant to the nature of a straight line." It is unlikely that he was satisfied with this, as he proceeds to a second attempt—also unsuccessful—and withheld publication of the book during his lifetime.

Despite the ultimate failure of Saccheri's labors, they did not go completely unrewarded. His probe did reveal a new replacement postulate of astounding simplicity: "there exists a Saccheri quadrilateral whose summit angles are right" (essentially 6C on p. 129), which reduces the problem to establishing the existence of a *single* figure.

It is hard to see how this new statement could be logically equivalent to the assertion that the summit angles of *every* Saccheri quadrilateral are right. Nonetheless it is, as I would now like to show. There are a couple of preliminaries.

Metatheorem[7] 8. *If in the context of Neutral geometry it were known that*

a given pair of straight lines have two common perpendiculars,

then it would follow that

all the perpendiculars drawn from one of the given straight lines, no matter how much it is produced, to the other (or the other produced) are equal (and thus the straight lines are equidistant).

Proof.
1. Let straight lines "*l*" and "*m*" have common   hypothesis
   perpendiculars "*PQ*" and "*ST*", and let "*XY*"
   be any other perpendicular drawn from *l* (or *l*
   produced) to *m* (or *m* produced) (see Figure 119).
2. *PQ = ST*   Th *A* (applied to
                                                      QTSP)

(We shall show *XY = ST*. This will suffice as *XY*, being *any* other perpendicular, represents *every* other perpendicular.)

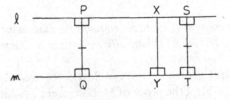

Figure 119

Case 1. *X* is between *P* and *S*.

3. Then *Y* is between *Q* and *T*.   exercise
4. Pretend *XY* < *ST*.   RAA hypothesis
5. ∠*SXY* > 90°   Th *A* (applied to
                                                      YTSX)
6. *XY* < *PQ*   2, 4, C.N. 6 (If *a* = *b*
                                                      and *c* < *b* then *c* < *a*.)
7. ∠*PXY* > 90°   Th *A* (applied to
                                                      QYXP)
8. Therefore ∠*PXY* + ∠*SXY* > 180°.   5, 7, C.N. 6 (If *a* > *b*
                                                      and *c* > *b* then
                                                      a + *c* > 2*b*.)
9. But ∠*PXY* + ∠*SXY* = 180°.   Th 13
10. Contradiction.   8 and 9
11. Therefore *XY* is not less than *ST*.   4–10, logic

(Similarly *XY* cannot be greater than *ST*, so in Case 1 *XY* = *ST*.)

Case 2. *X* is not between *P* and *S*, say *X* is on the other side of *S* than *P* (see Figure 120).

12. Then *Y* is on the other side of *T* than *Q*.   exercise
13. Pretend *XY* < *ST*.   RAA hypothesis
14. Produce *YX* and cut off "*Z*"*Y* = *ST*.   Post 2, Th 3
15. Draw *PZ* and *SZ*.   Post 1
16. ∠*ZST* = ∠*SZY*   14, Th *A* (applied to
                                                      TYZS)
17. ∠*SZY* < ∠*PZY*   C.N. 5
18. *PQ* = *ZY*   2, 14, C.N. 1
19. ∠*PZY* = ∠*ZPQ*   Th *A* (applied to
                                                      QYZP)
20. Therefore ∠*ZST* < ∠*ZPQ*.   16, 17, 19, C.N. 6
                                                      (If *a* = *b*, *b* < *c*, and
                                                      c = *d* then *a* < *d*.)
21. ∠*ZSX* > ∠*ZPS*   Th 16
22. ∠*XST* < ∠*SPQ*   20, 21, C.N. 6 (If
                                                      a < *b* and *c* > *d* then
                                                      a − *c* < *b* − *d*.)

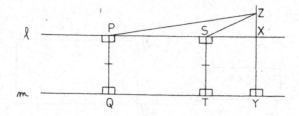

Figure 120

23. $\angle SPQ = \angle PST$ Post 4
24. $\angle XST < \angle PST$ 22, 23, C.N. 6 (If
                                                 $a < b$ and $b = c$ then
                                                 $a < c$.)

25. But $\angle XST = \angle PST$.   Post 4
26. Contradiction.   24 and 25
27. Therefore XY is not less than ST.   13–26, logic
(Similarly XY cannot be greater than ST, so in Case 2 $XY = ST$.)

Theorem C. *If the base of one Saccheri quadrilateral is equal to the base of a second Saccheri quadrilateral, and the equal sides of the first are equal to the equal sides of the second, then their summits and summit angles are also equal.*

The easy proof is an exercise.

Metatheorem 9. *If in the context of Neutral geometry it were known that*

there exists a Saccheri quadrilateral whose summit angles are right,

then it would follow that

the summit angles of every Saccheri quadrilateral are right.

Proof.
1. Let "$ABCD$" be a Saccheri quadrilateral whose   hypothesis
   summit angles are known to be right, and let
   "$WXYZ$" be any other Saccheri quadrilateral
   (see Figure 121).
2. Bisect DC at "M" and AB at "N", and draw MN.   Th 10, Post 1
3. MN is perpendicular to both DC and AB.   Corollary to Th A
4. $MN = CB$.   Metatheorem 8 (DC
                                                   and AB are the
                                                   straight lines, DA and
                                                   CB the common
                                                   perpendiculars)

Case 1. $MN < ZW$ (see Figure 122).
5. Produce NM and cut off "P"$N = ZW$.   Post 2, Th 3

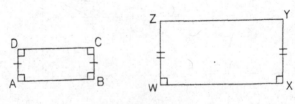

Figure 121

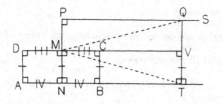

Figure 122

6. Through P draw $P``S"$ perpendicular to PN.   Th 11
7. From PS (or PS produced) cut off $P``Q" = $ (Post 2) Th 3
   WX.
8. Produce AB and draw $Q``T"$ perpendicular to it.   Post 2, Th 12
(In the figure I have drawn QT to the right of CB, but the proof will work perfectly well if QT is between PN and CB, or if the lower part of QT coincides with CB.)
9. Draw MQ and MT.   Post 1
10. DC produced will meet QT in a point V.[8]   Post 2, Post 6 (applied
                                                       to $\triangle QMT$), Post 1
11. $VT = CB$.   Metatheorem 8 (see
                  step 4)
12. $\angle CVT = 90°$ Th A (applied to
                          $BTVC$)
(We shall show that $QPNT$ is an upside-down Saccheri quadrilateral with summit NT, whose summit angles NTQ and TNP are equal to the summit angles of $WXYZ$.)
13. $PQ = NT$.   Metatheorem 8
                  (PN and QT are the
                  straight lines, VM and
                  TN the common
                  perpendiculars)
14. $\angle PQT = 90°$.   Th A (applied to
                           $PNTQ$)
15. $PN = QT$.   Th A (applied to
                  $NTQP$)
16. $QPNT$ can be considered a Saccheri quadri-   6, 14, 15, Def of
    lateral with summit NT.   "Saccheri quadrilateral"
17. The summit angles (NTQ and TNP) of Saccheri   5, 7, Th C
    quadrilateral $QPNT$ are equal to the summit
    angles (YZW and ZYX) of Saccheri quadri-
    lateral $WXYZ$.
18. The summit angles of Saccheri quadrilateral   8, 3, 17, Post 4
    $WXYZ$ are right.
(The proofs of Case 2 ($MN = ZW$) and Case 3 ($MN > ZW$) are only slight variations.)

Corollary 1. *The statement*

there exists a Saccheri quadrilateral whose summit angles are right

is logically equivalent to the statement

the summit angles of every Saccheri quadrilateral are right (replacement postulate 6A).

Proof. Neutral geometry + the first statement ⇒ 6A by Metatheorem 9. That Neutral geometry + 6A ⇒ the first statement is obvious, in view of the fact that Saccheri quadrilaterals are known to exist in Neutral geometry (p. 132).

Corollary 2. *The statement*

there exists a Saccheri quadrilateral whose summit angles are right

is logically equivalent to Postulate 5.

Proof. Combine Corollary 1 with Metatheorem 7's Corollary 1.

Corollary 3. *The statement*

there exists a triangle with angle-sum 180° (replacement postulate 7B)

is logically equivalent to Postulate 5.

Proof. There are two things to prove:

(1) Neutral geometry + 7B ⇒ Post 5, and
(2) Neutral geometry + Post 5 ⇒ 7B.

In view of the fact that triangles exist in Euclidean geometry, we proved (2) when we proved Theorem 32 (b). Here is a proof of (1).

1. There exists a triangle "*ABC*" with angle-sum   hypothesis
   180° (see Figure 123).
2. Bisect *AB* at "*D*" and *AC* at "*E*".   Th 10
3. Draw *DE* and produce it in both directions.   Post 1, Post 2
4. Draw *B*"*F*" and *C*"*G*" perpendicular to *DE*.   Th 12

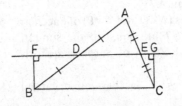

Figure 123

5. *GFBC* is a Saccheri quadrilateral with summit   Th B
   BC whose summit angles, *FBC* and *GCB*, have
   the same sum as the three interior angles of
   △*ABC*.
6. ∠*FBC* + ∠*GCB* = 180°   1, 5, C.N. 1
7. ∠*FBC* = ∠*GCB*   Corollary to Th A
8. Angles *FBC* and *GCB* are right.   6, 7, C.N. 6 (If $a +$
                                       $b = c$ and $a = b$ then
                                        $a = \frac{1}{2}c$ and $b = \frac{1}{2}c$.)
9. Therefore, there exists a Saccheri quadrilateral   5, 8
   whose summit angles are right.
10. Postulate 5 follows.   Corollary 2 (above)

An Experimental Test of Postulate 5

Either of the last two corollaries can be used as the basis for a simple scientific test of Postulate 5.

Corollary 2 would seem slightly preferable, as an experiment based on it would require the measurement of only one angle (either summit angle of the Saccheri quadrilateral). However, a few decades after Saccheri's book was published, the German-French mathematician, physicist, and philosopher Johann Heinrich Lambert (1728–1777) proved that, as figures get bigger, so do any discrepancies there may be between the actual sizes of their angles and the sizes predicted by Postulate 5-dependent theorems. Thus the bigger the figure, the more severely Postulate 5 is challenged. Big triangles are easier to construct than big Saccheri quadrilaterals, so actual experiments are usually based on Corollary 3.

In 1820 Gauss was placed in charge of a project to map Hannover, the Germanic state in which he lived. (The funds came in the name of George III of England, of American Revolution fame—from 1815 to 1837 the English kings ruled Hannover as well.) The project involved surveying huge triangles, using for vertices such things as steeples and mountains that could be seen despite the earth's curvature. To insure that the triangles were sited accurately Gauss invented the "heliotrope," a device whose finely adjustable mirror reflected sunlight in a direction the operator could control very precisely. Each triangle was measured several times and the slightly varying results averaged.

At the end of an 1827 paper on curved surfaces (*Disquisitiones generalis circa superficies curvas*), Gauss reports on the largest plane triangle to have been measured up to that time, with vertices on the mountaintops Hohenhagen, Brocken, and Inselsberg. (Figure 124 is a modern map of Germany with Gauss' triangle superimposed. Side BI is 107 km—66.5 miles—long!) His finding was that, within experimental error, the triangle's angle-sum was 180°.

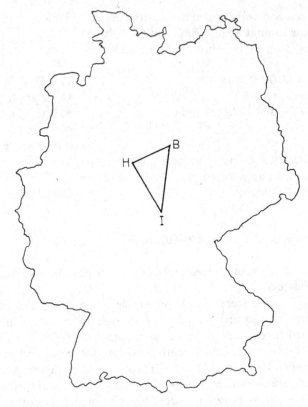

Figure 124

Gauss' only stated purpose in citing this result was to compare the straight-sided triangle HBI, which lies in a plane poised on the three mountaintops, to the corresponding curved-sided triangle drawn underneath it on the earth's surface.

Later mathematicians have suggested that these measurements also had the additional goal of investigating whether or not the triangle HBI, constructed of light rays, had a sum of angles that deviated from the Euclidean value of 180°.

The paper on curved surfaces, which to be sure concerned only Euclidean geometry, offers no support for such a suggestion. But in the letter to Taurinus [dated November 8, 1824] ... for example, there is a hint in that direction. It is not absolutely out of the question that Gauss, by using his great triangle, attempted to find empirically the deviation from Euclidean geometry in the space of the universe .... [F]rom his own remarks we know that he considered geometry an empirical science in the same class as mechanics.*

---

* Tord Hall: *Carl Friedrich Gauss*, MIT Press (p. 124). Copyright © 1970 by MIT Press. Reprinted with permission.

In any case, Gauss' measurement of the giant triangle does constitute experimental support of Postulate 5, whether he viewed it so or not.

Does this mean that Postulate 5 has finally been proven? Not at all. Mathematics and experimental science have quite different standards.

Consider, for example, the notion of length. In mathematics lengths are fixed and exact. We even classify them as rational or irrational (p. 3). But no instrument can determine the length of a physical object to more than a few decimal places. The result of a measurement, therefore, is never an exact number but rather a *range* of numbers consisting of infinitely many individual numbers that begin with the determined digits. (This is obscured by the fact that when the range is narrow enough for practical purposes, we tend to describe one convenient number within it as the "exact" length.) Thus a measured length is *not* a length in the mathematical sense, and in experimental science fine mathematical distinctions like rational/irrational are meaningless. (In fact, if we ignore instrumental limitations and compare mathematics to *theoretical* science, the mathematical notion of length is still without a counterpart. According to current theory, every physical object—even a ray of light—consists of erratic particles and fluttering waves whose precise locations are necessarily indeterminate, even under the hypothesis of ideal instruments. In particular, then, it is not permissible to speak of an object as having a well-defined "beginning" or "end," or therefore a "length" in the mathematical sense.)

Gauss' measurement of the giant triangle does not prove Postulate 5 because his finding—that the angle-sum was within a certain experimental error of 180°—means only that the "true" value was in a narrow but infinitely-populated range that happened to include the conveniently round number 180. Indeed, *no* measurement of a triangle could possibly prove Postulate 5, for however much new techniques might shrink the range of experimental error, it is not conceivable that they could eliminate it entirely.

My point is important so let me make it again, in a different way. I imagine before me an impeccably flat surface. Using a perfect straightedge, an exquisitely made compass, and a pencil of utmost sharpness, I draw a right triangle, *ABC* in Figure 125, and a straight line *AD* making ∠*DAC* = ∠*ACB*. The angle-sum of triangle *ABC* will then be equal, or not, to 180° according as ∠*DAB* is, or is not, a right angle. To see if ∠*DAB* = 90° I draw a straight line *AE* perpendicular to *AB*. *AE* appears to fall right on top of *AD*! Even through

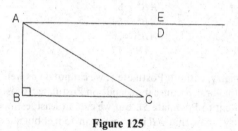

Figure 125

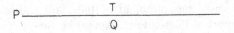

Figure 126

a powerful microscope I can detect no divergence between AE and AD. Doesn't that prove Postulate 5?

For practical purposes, perhaps; to a mathematician's satisfaction, definitely not. Consider for instance a straight line PQ (Figure 126). Produce PQ until its length is ten billion light-years. At its far end draw a perpendicular 0.0000000001 cm long (a length much smaller than the wavelength of visible light). From the top of this tiny perpendicular draw a straight line back to P. On this last straight line choose a point T making $PT = PQ$. Then, regardless of the delicacy of the tools used, or the care with which the figure is examined, the straight lines PT and PQ will be indistinguishable (in fact throughout the entire 10,000,000,000 light-year length) because the space between them is simply too narrow to be seen. It is possible then that the straight lines AE and AD in Figure 125 are actually distinct and Postulate 5 false. (Exactness is a standard that allows no moderation.)

It is ironic that though no measurement of a triangle can prove Postulate 5, it is possible to conceive of a measurement that would disprove it. Suppose for example that at some time in the future, using instruments much more sensitive than any we have today, the angle-sum of a triangle is found to be 179.9999999972°, with an experimental error of plus or minus 0.0000000001°; then the true value would be in the range of 179.9999999971° to 179.9999999973°, and so definitely less than 180°.

Exercises

1. Let ABC (Figure 127) be a right-angled triangle in which the sides containing the right angle are 8 and 15 feet long. In Euclidean geometry we would have the Theorem of Pythagoras (Theorem 47) available and so be able to calculate AB exactly:

$$AB^2 = AC^2 + BC^2$$
$$AB^2 = 8^2 + 15^2$$
$$AB^2 = 64 + 225$$
$$AB^2 = 289$$
$$AB = 17.$$

In Neutral geometry, without Postulate 5, we cannot do as well. We don't have the Theorem of Pythagoras because it depends on Postulate 5 (it is, as a matter of fact, logically equivalent to Postulate 5). But we can, at least, estimate AB. Using only Neutral geometry, show that AB is longer than 15 feet but shorter than 23 feet.

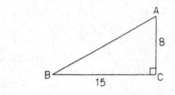

Figure 127

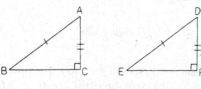

Figure 128

2. The "hypotenuse" of a right-angled triangle is the side subtending the right angle. Using only Neutral geometry, prove the Hypotenuse-Side Theorm:

   Theorem (HS). *If the hypotenuse and another side of one right-angled triangle are equal, respectively, to the hypotenuse and another side of another right-angled triangle, then the triangles are congruent* (see Figure 128).

   Hint: For an easy proof by contradiction, pretend $\angle BAC \neq \angle EDF$. For a more interesting direct proof, produce BC to "G" so that $CG > EF$.

3. There's no SSA theorem because SSA will not, in general, insure congruence. (Exercise 2 was a special case.) In Figure 129, for example, where $\triangle ABC$ is equilateral, triangles ADB and ADC are not congruent, even though two sides and an angle are equal: $AB = AC$, $AD = AD$, $\angle ADB = \angle ADC$. With one extra condition, though, a theorem (I'll call it "SSA +") is possible; and like all the other congruence theorems, it is part of Neutral geometry.

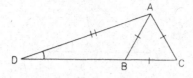

Figure 129

   Theorem (SSA +). *If two sides of one triangle are equal respectively to two sides of another triangle, and the angles subtended by one pair of equal sides are equal, then, if the angles subtended by the other pair of equal sides are*

   (1) *both acute, or*
   (2) *both obtuse, or*
   (3) *if one of them is right,*

   then the triangles are congruent (see Figure 130).

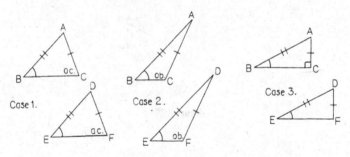

Figure 130

Prove this theorem, using only Neutral geometry. The three cases should be taken up individually. In each case aim to show that $\angle BAC = \angle EDF$.

4. Here's what looks like an uncomplicated proof, using only Neutral geometry, that the angle-sum of every triangle is 180°. We should be suspicious of it, though. That the angle-sum of every triangle is 180° is logically equivalent to Postulate 5 (Corollary 2 to Metatheorem 7, p. 141), making this simple argument, if correct, a proof of Postulate 5!

Find the unstated assumption, and prove it is logically equivalent to Postulate 5.

Let's take the attitude that we don't know *what* the angle-sum of a triangle is. We'll call it "x." Let "ABC" be any triangle. Choose "D" at random between B and C, and draw AD. Label the angles as in Figure 131. Now $\angle 1 + \angle 2 + \angle 3 = x$, because we're letting "x" stand for the angle-sum of a triangle. Similarly $\angle 4 + \angle 5 + \angle 6 = x$. Adding those two equations together we get $\angle 1 + \angle 2 + \angle 3 + \angle 4 + \angle 5 + \angle 6 = 2x$. But $\angle 1 + \angle 2 + \angle 5 + \angle 6$ is also the angle-sum of a triangle, $\triangle ABC$, so $\angle 1 + \angle 2 + \angle 5 + \angle 6 = x$. Subtracting this equation from the previous one gives $\angle 3 + \angle 4 = x$. But $\angle 3 + \angle 4 = 180°$ by Theorem 13, so $x = 180°$. Thus the angle-sum of a triangle is 180°.

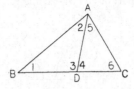

Figure 131

5. Prove that Playfair's Postulate (3C on page 128) is logically equivalent to Postulate 5.

Note: As Poseidonios' Postulate wasn't used explicitly in the proof of Theorem P2 (page 122), that proof can be viewed as a proof that

Neutral geometry + Playfair's Postulate ⇒ Postulate 5.

(Playfair's Postulate, in the guise of "Theorem P1," *was* used in the proof, in step 7.) Thus all that remains to show is that

Neutral geometry + Postulate 5 ⇒ Playfair's Postulate.

6. Show that Lorenz's Postulate (9A on page 129) is logically equivalent to Playfair's Postulate (and therefore, by Exercise 5, to Postulate 5).

Notes

[1] *Poseidonios* was head of the Stoic school on Rhodes and a teacher of Cicero. His works have not survived, but some of their titles (which have) indicate the breadth of his interests: *Treatise on Ethics*; *Treatise on Physics*; *History of Pompey's Campaigns in the East*; *On the Universe*; *Introduction to Diction*; *On Emotions*; *Against Zeno of Sidon* (on geometry).

[2] *see for ourselves*. We will establish shortly that Poseidonios' Postulate and Euclid's Postulate 5 are but different formulations of the same basic assumption. Some scholars feel that Euclid was aware of the formulation later used by Poseidonios but chose the formulation he did precisely because of its finite character.

[3] *Neutral geometry*. The term seems to have originated with Prenowitz and Jordan in *Basic Concepts of Geometry* (Blaisdell, 1965). Most books use the term "Absolute geometry" introduced in 1832 by János Bolyai, one of the founders of non-Euclidean geometry.

[4] *Wallis' Postulate*. Wallis actually made the stronger assumption that "to *every* figure there exists a similar figure of arbitrary magnitude." (Bonola, *Non-Euclidean Geometry* (1906; Dover reprint, 1955), pp. 15–17.)

[5] *letter*. From Bonola, *op. cit.*, pp. 65–66.

[6] *the following*. For technical reasons geometers classify this law as a "continuity" axiom, like Postulates 6 and 7, even though it does not appear to assert continuity. Unlike Postulates 6 and 7, it is explicitly stated in the *Elements*, though not in Book I: it is part of Definition 4 at the beginning of Book V. It is now known as the "Axiom of Archimedes" because, a few decades after the *Elements* appeared, Archimedes pointed out that Euclid wielded it more like an axiom than a definition.

[7] *Metatheorem*. The distinction between "theorem" and "metatheorem" is largely in the eye of the beholder. So far in this article I've called a result a "theorem" (Theorem A, p. 133 and Theorem B, p. 134) when its hypothesis was known to be possible in Neutral geometry, and a "metatheorem" when the possibility of its hypothesis had not been established. Since a conditional statement does not assert the hypothesis, but merely that it entails the conclusion, this pattern is not binding. Nonetheless I will follow it; thus this is a "metatheorem," even though it has the ring of a "theorem."

[8] *DC produced will meet QT in a point V*. I know DC enters $\triangle QMT$ because, by Theorem 27/28, $DC \| PS$ and $DC \| AT$ (in each case PN is the third straight line).

CHAPTER 5

The Possibility of Non-Euclidean Geometry

By the latter half of the 18th century the problem of deducing Postulate 5 from Neutral geometry had become, in mathematical circles, notorious. The Encyclopedist Jean le Rond d'Alembert called it (1759) "le scandale des éléments de géométrie."

It was only a matter of time before the great difficulty of the problem would cause some to conclude that its solution was impossible. Apparently the first to do so in print was G. S. Klügel (1739–1812), a doctoral student at the University of Göttingen, with the support of his teacher A. G. Kästner, in the former's 1763 dissertation *Conatuum praecipuorum theoriam parallelarum demonstrandi recensio* (Review of the Most Celebrated Attempts at Demonstrating the Theory of Parallels). In this work Klügel examined 28 attempts to prove Postulate 5 (including Saccheri's), found them all deficient, and offered the opinion that Postulate 5 is unprovable and is supported solely by the judgment of our senses.

Of course that was only Klügel's opinion; he could not *prove* that Postulate 5 is unprovable. Nonetheless the idea was taken seriously by others, and it turned out to be the appeal of precisely this unproven idea that made the eventual discovery of non-Euclidean geometry inevitable. It is but a tiny transition (logically) from speculating that

(1) Neutral geometry by itself does not imply Postulate 5,

to speculating that

(2) a new geometry contrary to Euclid's is logically possible.

If you don't see the connection between (1) and (2), or are confused by what you do see, don't judge yourself too harshly. The *psychological* distance between the two statements is enormous. The first mathematicians to pass from (1) to (2) did not do so until *fifty* years later!

This chapter is about the passage from (1) to (2), and the meaning of (2).

We will begin by assuming that statement (1) is correct. Klügel concluded that it was, based on his survey of the attempts to prove Postulate 5, and I think you will agree, after our own survey in Chapter 4, that statement (1) is at least plausible.

Statement (1) can be proven; in fact we will describe such a proof in Chapter 7. To do so now, however, seems a needless complication. When the men who first made the transition from (1) to (2) did so, they had not seen a proof of (1) either.[1]

In the next section we will make our own transition from (1) to (2). Once some background from modern logic has been sketched in, this will be a matter of only a few steps. But since *taking* those few steps will be enormously difficult if we realize where they are leading us, we will endeavor to take them inattentively, even recklessly, unmindful of what the meaning of (2) might be.

Then, in the remainder of the chapter, committed by logic to (2), we will clamber for grounds on which we can at least provisionally accept the unimaginable state of affairs in which we find ourselves.

The Logical Possibility of Non-Euclidean Geometry

Implicit in the study of any axiomatic system is the assumption that its axioms are *consistent*, i.e., that from them it is impossible to deduce a contradition.[2] We assume consistency, for example, whenever we construct a proof by contradiction: what enables us to attribute sole responsibility for our contradiction to the RAA hypothesis is our belief that no contradictions would have been possible without it. But the importance of consistency is not confined to its bearing on a single method of proof. Without consistent axioms it is impossible to deduce a single significant theorem by any technique whatever, because the presence of a contradiction violates one of logic's basic laws (the Law of Contradiction, p. 13) and thereby short-circuits the operation of logic itself.[3]

Thus the very logical possibility of an axiomatic system requires that the axioms be consistent.

It so happens that the logical possibility of an axiomatic system doesn't require anything else. Logic is concerned with the *relations among* statements—Are they compatible? Does this one follow from the others?— but *not* with their substance. The statements involved in a deductive process may strike us as false, or even nonsensical—

All purple flamingoes have seven toes on each foot; the present Governor of Massachusetts is a purple flamingo; therefore, the present Governor of Massachusetts has seven toes on each foot

—but the argument will nonetheless be logical as long as its *form* is valid. The *worth* of the deduction is simply not a logical concern. Thus we see that unless it is blocked at the outset by inconsistency, logic is free to operate on *any* collection of premises and develop it into an axiomatic system.

The foundation of the "new geometry" entailed by the unprovability of

Postulate 5 is made up of the foundation of Neutral geometry and the negation of Postulate 5 (which I will denote by "~ Postulate 5"). We have just seen that the logical possibility of an axiomatic system resides entirely in the consistency of its foundation; therefore, to establish that the new geometry is logically possible—this is statement (2)—we have only to show that no contradictions are deducible from

Neutral foundation $\begin{cases} \text{Euclid's primitive terms and Definitions of defined} \\ \text{terms, Euclid's Common Notions, Postulates 1–4,} \\ \text{Postulates 6–10, and} \end{cases}$
$\qquad\qquad$ ~ Postulate 5.

Before we do that let me pause for a few remarks.

First, the content of ~ Postulate 5 is of course that Postulate 5 is false, but we are for the moment doing ourselves a favor by not reflecting on what that might mean (especially by not picturing what it would look like!).

Second, if you do allow yourself to feel some of your commitment to Postulate 5, you may wonder how we can hope to show the new geometry's foundation to be consistent when it contains a component that is obviously (to you) false. If you are in this frame of mind please remember that consistency is an *internal* property of a set of premises—it is not a matter of their truth (relation to the world) but only of their compatibility (relation to each other), and that true and false premises can be compatible if they and their offspring manage to avoid a logical head-on collision.

Finally, I want to emphasize the fundamental divergence between what we are doing now and what we were doing in Chapter 4, so that you will not be misled by their surface similarity. In Chapter 4 we took Euclidean geometry, whose foundation is

$$\text{the Neutral foundation} + \text{Postulate 5,}$$

and replaced Postulate 5 by various statements (such as Poseidonios' Postulate) that turned out to be *logically equivalent* to it. Thus in each case the system we studied was still Euclidean geometry, albeit organized on a different foundation. But what we are doing now is drastically different. In the foundation we are considering now—

$$\text{the Neutral foundation} + \text{~ Postulate 5}$$

—the Euclidean postulate has been replaced by its *direct opposite*.

Back to the proof that statement (1) implies statement (2). We said a few paragraphs ago that all we have to show is that no contradictions can be deduced from

Euclid's primitive terms and Definitions of defined terms, Euclid's Common Notions, Postulates 1–4, Postulates 6–10, and ~ Postulate 5.

This can be established at once. Just notice that if a contradiction were deducible from these premises, then the very same deduction could be viewed, from another angle, as a proof by contradiction of Postulate 5!

I'd better elaborate. Our hypothesis is statement (1), by which no deduction of Postulate 5 from Neutral geometry is possible. In particular, then, no *proof by contradiction* of Postulate 5 from Neutral geometry is possible. Such a proof would schematically look like this:

Schema I. Start with Euclid's primitive terms, Definitions of defined terms, Common Notions, Postulates 1–4, and Postulates 6–10; pretend, as an RAA hypothesis, that Postulate 5 is false; deduce a contradiction.

This then is impossible.

Our desired conclusion is statement (2), which we have reduced to the assertion that the following deduction is impossible:

Schema II. Start with Euclid's primitive terms, Definitions of defined terms, Common Notions, Postulates 1–4, Postulates 6–10, and ∼Postulate 5; deduce a contradiction.

The only difference between the two is that in schema I ∼Postulate 5 is singled out for only provisional acceptance and in schema II it is not. It amounts to a difference of purpose. Someone seriously attempting to carry out schema I would be trying to prove Postulate 5, whereas someone working under schema II would be trying to show that the proposed new geometry is inconsistent.

Of course differences of intent are outside the purview of logic. Logically, the two schemata are identical. Thus the impossibility of schema I, which was given, coincides with the impossibility of schema II, which was to be proved.

The Founders of Non-Euclidean Geometry

[T]here is some truth in this, that many things have an epoch, in which they are discovered at the same time in several places, just as violets appear on all sides in springtime.

—letter from Farkas Bolyai to his son János
(Bonola's *Non-Euclidean Geometry*, p. 99)

Credit for discovering the new geometry belongs to those thinkers who believed in at least the possibility of statement (1) and, without knowledge of others' work, came to recognize that it entailed statement (2). In this sense it appears that non-Euclidean geometry was discovered no less than four times in a twenty-year period. Multiple independent discoveries are not uncommon in the histories of science and mathematics, especially at times when a number of people are working on the same problem and communication among them is scanty.

Gauss seems to have been the first, around 1813, to have a clear view of a consistent geometry in which Postulate 5 is replaced by its negation. This came to him after twenty years of sporadic attempts to prove Postulate 5 (see Gauss's letter on p. 127). During the next few years he investigated the new geometry—it was he who eventually called it "non-Euclidean"—and uncovered a number of its theorems.

Then, in late 1818 or early 1819, Gauss received a note from a professor of jurisprudence named Ferdinand Schweikart (1780–1859) indicating that Schweikart had independently come to basically the same conclusions.

Had either Gauss or Schweikart published their discovery, non-Euclidean geometry would perhaps be regarded today as having had only two founders. This is particularly likely in the case of Gauss, whose reputation was international. But neither one did. As far as I know the reasons for Schweikart's reticence are unknown, but from Gauss's letters we learn that his reasons were fear of the "clamor"[4] he felt would be set off by the inevitable misunderstanding of the discovery, and "a great antipathy against being drawn into any sort of polemic."[5]

Early in 1831, though still of a mind "not to allow it to be published during my lifetime,"[6] Gauss set out to record what he had learned of non-Euclidean geometry.

[This] I had never put in writing, so that I have been compelled three or four times to go over the whole matter afresh in my head. But I wished that it should not perish with me.[7]

But less than a year later Gauss received from Farkas Bolyai a copy of a treatise on non-Euclidean geometry soon to be published by Farkas's son János Bolyai (1802–1860). With the future of non-Euclidean geometry thus assured, an apparently relieved Gauss broke off his own project.

[I]t is . . . a pleasant surprise for me that I am spared this trouble, and I am very glad that it is just the son of my old friend, who takes the precedence of me in such a remarkable manner.[8]

János Bolyai's *Absolute Science of Space* was published as an appendix to a mathematical book of his father's later that year (1832).

The fourth founder of non-Euclidean geometry was Nicolai Ivanovich Lobachevsky (1793–1856), professor of mathematics at the University of Kazan, Russia. While he actually broke into print before János Bolyai, publishing a paper on non-Euclidean geometry (*On the Principles of Geometry*) in 1830, his thoughts were still directed toward proving Postulate 5 as late as 1823, at which time the basic idea of non-Euclidean geometry was already in János Bolyai's possession. Lobachevsky's early papers on non-Euclidean geometry were written in Russian; his work did not become known in central or western Europe (the Bolyais were Hungarian, Gauss German) until some years later, when he began to publish accounts in French (1837) and German (1840).

Beginning with an epochal paper delivered by George F. B. Riemann (1826–1866) at the University of Göttingen in 1854, mathematicians realized that replacing Postulate 5 with its negation was not the *only* way Euclidean geometry could be tampered with, and within a few years other consistent non-Euclidean geometries made their appearance. We will not take them up in this book, but in view of their existence the non-Euclidean geometry of Gauss, Schweikart, János Bolyai, and Lobachevsky is really only the *first* non-Euclidean geometry. It is usually known today as "hyperbolic geometry," a name suggested by the mathematician Felix Klein (1849–1925) in 1871. In Greek *hyperbole* means "excess," and in the geometry of Gauss, Schweikart, János Bolyai, and Lobachevsky the number of parallels through a given point to a given straight line is in *excess* of the number in Euclidean geometry.

The Psychological Impossibility of Non-Euclidean Geometry

Postulate 5 is logically equivalent to Playfair's Postulate:

Through a given point not on a given straight line, and not on that straight line produced, no more than one parallel straight line can be drawn.

(This is replacement postulate 3C on p. 128. If you happened to try Exercise 5 on p. 152, you may be already in possession of a proof that Postulate 5 and Playfair's Postulate are logically equivalent.[9] If not, there's a proof on p. 172.) Thus ∼Postulate 5 is logically equivalent to ∼Playfair's Postulate (the negation of Playfair's Postulate). Since the meaning of Playfair's Postulate is that for *every* combination of a point and a straight line—the point not on the straight line or the straight line produced—no more than one parallel can be drawn, its negation will assert the existence of an exception:

∼**Playfair's Postulate.** There exists at least one point P and at least one straight line AB such that

(1) P is not on AB or AB produced, and
(2) through P there are at least two straight lines parallel to AB.

In hyperbolic geometry it is customary to use this version of ∼Postulate 5 instead of ∼Postulate 5 itself.

You may feel at this juncture that you don't understand.

"Excuse me," I imagine someone saying, "I thought I heard you say that in hyperbolic geometry we accept Euclid's primitive terms, Definitions of defined terms, Common Notions, Postulates except #5—"

That's right. We accept Neutral geometry.

"—and that we replace Postulate 5 with ∼Playfair's Postulate."

Yes. ∼Playfair's Postulate is only ∼Postulate 5 in different words.

"But how can there be two parallels to the same straight line through the same point? Look [drawing Figure 132]. Here's point P, here's line AB. I can draw a parallel, say CD, through P—"

Yes, by Euclid's Theorem 31, which is part of Neutral geometry.

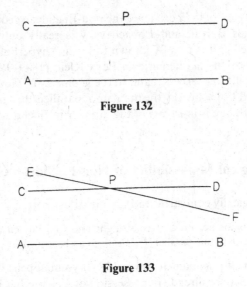

Figure 132

Figure 133

"But now, how can I draw another one? Do you mean the two parallels are right on top of one another?"

No, they are distinct straight lines, having no points in common except *P*.

"Oh, I get it! The other parallel comes out of the page, into the third dimension!"

No. There *is* a three-dimensional hyperbolic geometry, just as there's a three-dimensional Euclidean geometry. But this book is concerned only with *plane* geometry. The other parallel is in the same plane as *CD* and *AB*.

"But then [drawing Figure 133] it would look like this!"

Angle *DPF* may not be as large as you've drawn it, but that's the idea, yes. [Silence. Then,] "Wait. Parallel lines are straight lines that never meet."

Yes. That's Definition 23.

"No matter *how much* they are produced."

Yes. In either direction.

"Then *EF can't* be parallel to *AB*!"

Why not?

[Suspicious silence.]

This isn't a trick, honestly. The picture looks strange to me, too, or rather, it seems to contradict what we are saying about it. But let's talk about why, *apart* from the appearance of the drawing, you think *EF* produced will hit *AB* produced. Can you *prove* that it will?

"Don't we have a theorem that says straight lines parallel to the same straight line are parallel to one another? If *CD* and *EF* are *both* parallel to *AB* they'll be parallel to one another, but they're not because they meet at *P*."

We *did* have that theorem, but not anymore. It's Theorem 30. We proved it with Postulate 5. In fact it's logically equivalent to Postulate 5.

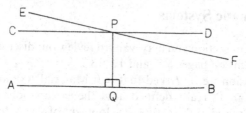

Figure 134

"... Let's get some angles. Draw the perpendicular [Figure 134] from *P* to *AB*. Call the foot *Q*. Then angle *DPQ* is a right angle."

It is if *CD* was drawn by the method used in the proof of Theorem 31. Let's say it was.

"OK, then *DPQ* is a right angle. And *FPQ* is smaller by Common Notion 5, so—"

May I interrupt? Were you about to say that angles *FPQ* and *PQB* add up to less than 180° and so *EF* and *AB* meet?

[Sighing] "—by Postulate 5. Yes, I was. We don't have Postulate 5."

[Smugly] Right.

[Time goes by. More proofs are suggested. Trudeau, with obvious pleasure, demonstrates how each proposal turns on some assertion logically equivalent to Postulate 5. Finally the other lapses into exasperated silence.]

[Conciliatory] I know. It is *so* clear from the drawing that *EF* meets *AB* that you can't help but think it would be easy to prove. You're not foolish for trying, and I'm sorry if I got carried away and made you feel that you were. On the contrary, you're in excellent company. Do you realize you've been wrestling with the very problem that tantalized some of the best minds in the world for two thousand years? You've been trying to prove that *CD* is the only parallel, which is what Playfair's Postulate says. You've been trying to prove Postulate 5!

[Suddenly exhausted] "My God, I have."

But in this chapter we've agreed to assume that Postulate 5 is unprovable.

"Yes. But ..."

But?

"OK, so maybe I can't prove that *EF* meets *AB*. If mathematicians couldn't do it in two thousand years—and I see now that I *was* trying to do just what they had tried to do—it's not likely that I'd succeed in an hour. But that doesn't make me believe for a minute that *EF* won't meet *AB*. *EF* has to hit *AB*. It's a straight line. It *has* to."

Are you saying that ~ Playfair's Postulate is "repugnant to the nature of a straight line"?

"Yes, I guess I am."

Saccheri said the same thing once, in a similar situation (page 142).

"Well, you said I'm in good company."

Formal Axiomatic Systems

Before reading this section you may want to review our discussion of Material Axiomatic Systems on pages 5–7 and 12–13.

In that discussion we remarked that in a Material Axiomatic System the primitive terms aren't really defined; that the explanations given in lieu of definitions aren't used in the actual development of the system; and that the only properties of primitive terms that *are* used are contained in the axioms.

Later this was borne out in our study of Euclidean geometry. Neither the explanation of the primitive term "point"—"that which has no part"—nor that of the primitive term "straight line"—"breadthless length which lies evenly with the points on itself"—nor that of any other primitive term was ever used as the reason for a step in the proof of a theorem.

This is not to say that the explanations of primitive terms were useless to us. On the contrary, they helped us to visualize what geometry was about, to design figures, to intuit what could be proved, and in general to see *meaning* in the Euclidean system. But these contributions pertain to *our* relationship *to* Euclidean geometry, and only to the relationship's imaginative and emotional aspects at that. The explanations of primitive terms had no role in the formal execution of Euclidean geometry, nor have they one in maintaining its objective existence as an artifact that can be studied on a logical level.

Explanations of primitive terms remind me of the guidebooks tourists read while strolling around the Great Pyramid. By indicating the plan of the monument and its composition, by speculating on its purpose and construction, they mitigate its strangeness, facilitate a relationship between it and the onlooker, and thereby make the contemplation of its great abstract bulk a more meaningful experience. But if there were no guidebooks, the monument would still stand, and could still be apprehended.

During the nineteenth century mathematicians' view of their subject was shaken by a *number* of unexpected developments, of which the discovery of non-Euclidean geometries was only the most startling. It was while struggling to assimilate the meaning of this multitude of unsettling events that they came, for the first time, to appreciate the fundamental significance for mathematics of what we have just been discussing, something they had been tacitly aware of all along: that the existence of an axiomatic system, considered apart from people's interpretations of it, does not depend on the explanations of its primitive terms. Boldly they made this realization the keystone of a radical new view of mathematical systems (compare "Pattern for a Material Axiomatic System", p. 6).

Pattern for a Formal Axiomatic System[10]

1. Certain undefined technical terms are introduced. These terms, called *primitive terms*, are the basic terms of the discourse.

2. A list of unsupported statements about the primitive terms is given and accepted. These primary statements are called *axioms*.
3. All other technical terms (the *defined terms*) are defined by means of previously introduced terms.
4. All other statements of the discourse are logically deduced from previously accepted or established statements. These derived statements are called *theorems*.

The departures from the earlier pattern are in items 1 and 2. Where the pattern for a Material Axiomatic System stipulated that there be explanations of the primitive terms, and that the axioms' acceptability be judged in light of this information—conditions dictated by the natural desires of human participants—the new pattern confines itself strictly to what is required to enable the system to function logically. Item 1: there must be primitive terms. They cannot be defined within the system, therefore within the system they are totally without meaning. They are names, no more, for the different types of object the system deals with. Item 2: there must be axioms. As they involve the primitive terms they also are without meaning, and therefore are neither true nor false. The acceptance we grant them is no more than our agreement to work out their consequences.

A Formal Axiomatic System is like a Material Axiomatic System seen through a fluoroscope. Meaning and truth have melted away, leaving just the logical skeleton.

Formerly the essence of a mathematical system was considered to lie in the *combination* of its logical skeleton and the meaning attached thereto; the modern view is that a mathematical system is, at root, *only* a logical skeleton, to which meaning may or may not be attached. The distinction is philosophically subtle, so it is fortunate that we will not need to concern ourselves with it to any great degree.

As I tried to suggest in the dialogue at the end of the last section, the confusion most people feel when they first encounter hyperbolic geometry swirls around the fact that they find it simply unimaginable. When the average person, looking at Figure 134, even *begins* to consider the possibility that *CD* and *EF* might both be parallel to *AB*, the person's intuition and imagination erupt in such a pandemonium of objections that the thought is lost before fully formed.

The trick is to lull these parts of the mind by regarding hyperbolic geometry as a Formal Axiomatic System. We hereby adopt this perspective until further notice. To so view Neutral geometry—which forms a part of hyperbolic geometry—is not to deny, but merely to ignore, the meaning we have attached to it; while to so view the rest of hyperbolic geometry is only to recognize that we have no other choice, that initially it *is* unimaginable. But if we admit hyperbolic geometry's lack of intuitive appeal right up front, say that in fact, being a Formal Axiomatic System, it isn't *supposed* to make much sense, our hope is that we can buy enough peace and quiet to be able to prove some of its theorems. Just getting the logical lay of the land will dilute the new geometry's strangeness, and carry us a long way toward eventual reconciliation with our intuitions and imaginations.

A Simple Example of a Formal Axiomatic System

Yes, now I remember, yesterday evening we spent blathering about nothing in particular.

—Samuel Beckett's *Waiting for Godot*, Act II

While you may comprehend intellectually what a Formal Axiomatic System is, you probably don't have much feeling for what it would be like to work with one. And until you do the parts of your mind we want to pacify won't shut up. Thus this example.

Figure 135

The Scorpling Flugs[11]

primitive terms: *scorple, flug*

axioms: SF1. If *A* and *B* are distinct flugs, then *A* scorples *B* or *B* scorples *A* (the possibility of both happening is not excluded).

SF2. No flug scorples itself.

SF3. If *A*, *B*, and *C* are flugs such that *A* scorples *B* and *B* scorples *C*, then *A* scorples *C*.

SF4. There are exactly four flugs.

Referring to Formal Axiomatic Systems, the English philosopher, mathematician, and social reformer Bertrand Russell (1872–1970) once said, "Mathematics is the subject in which we never know what we are talking about, nor whether what we are saying is true." We don't know what we're talking about because the primitive terms are meaningless, and all other terms refer back to them. And we don't know whether what we're saying is true because we don't even know what it means.

Nonetheless we can prove theorems. And since scorpling is apparently some sort of relation between flugs (whatever they are) we can even illustrate the theorems, though only very abstractly. Eventually, as our figures incorporate more and more information and we find ourselves able to anticipate new theorems, we may begin to suspect that not knowing what we're talking about is no great handicap after all.

Theorem SF1. *If a flug scorples another, it is not also scorpled by the other.*

Proof.

1. Let flug "*A*" scorple flug "*B*".   hypothesis
2. Pretend that *A* is also scorpled by *B*, i.e. that *B*   RAA hypothesis
   scorples *A*.
3. Then *A* scorples *A*.   1, 2, Ax SF3 (*A*, *B*, *A*
                                                                the three flugs)
4. But *A* does not scorple *A*.   Ax SF2
5. Contradiction.   3 and 4
6. Therefore *B* does not scorple *A*.   2–5, logic

Thus the possibility left open in Axiom SF1 is ruled out after all.

Corollary. *Given two distinct flugs, either the first scorples the second or the second scorples the first, but not both.*

Proof. Combine Axiom SF1 and Theorem SF1.

Theorem SF2. *If A scorples B and C is distinct from A, then A scorples C or C scorples B (possibly both).*

Proof. (The theorem can be restated as follows: If *A* scorples *B* and *C* is distinct from *A* and *A* does not scorple *C*, then *C* scorples *B*. This is the version we will prove.)

1. *C* scorples *A* or *A* scorples *C*, not both.   hypothesis (*C* distinct
                                                       from *A*), Corollary
2. *A* does not scorple *C*   hypothesis
3. *C* scorples *A*   1, 2
4. *A* scorples *B*   hypothesis
5. *C* scorples *B*   3, 4, Ax SF3 (*C*, *A*, *B*
                                                       the 3 flugs)

Theorem SF3. *There is at least one flug that scorples every other flug.*

Figure 136

Figure 137

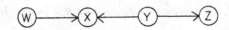

Figure 138

Proof.

1. Let "*W*", "*X*", "*Y*", "*Z*" be the four distinct   Ax SF4
   flugs.
2. Either *W* scorples *X* or *X* scorples *W*, but not   Corollary
   both; to be specific, say *W* scorples *X*.
3. Similarly we can say *Y* scorples *Z* (Figure 136).   Corollary
4. *W* scorples *Y* or *Y* scorples *X*, or both.   2, 1 (*Y* ≠ *W*), Th SF2
      (*W*, *X*, *Y* the 3 flugs)

(Thus there are three cases to consider.)

Case 1. *W* scorples *Y* but *Y* does not scorple *X* (Figure 137).

5. Then *W* scorples *Z*.   hypothesis of Case 1
      (*W* scorples *Y*), 3,
      Ax SF3 (*W*, *Y*, *Z* the
      3 flugs)

6. *W* scorples every other flug.   2, hypothesis of Case
      1, 5

Case 2. *Y* scorples *X* but *W* does not scorple *Y* (Figure 138).

7. *W* scorples *Y* or *Y* scorples *W*, not both.   Corollary
8. *W* does not scorple *Y*   hypothesis of Case 2
9. *Y* scorples *W*   7, 8
10. *Y* scorples every other flug.   9, hypothesis of Case
       2, 3

Case 3. *W* scorples *Y* and *Y* scorples *X* (Figure 139).

11. *W* scorples *Z*   hypothesis of Case 3
       (*W* scorples *Y*), 3, Ax
       SF3 (*W*, *Y*, *Z* the
       3 flugs)

12. *W* scorples every other flug.   2, hypothesis of Case
       3, 11

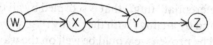

Figure 139

Definition SF1. A flug that scorples every other flug is called a *pushy* flug.

Theorem SF4. *There is one and only one pushy flug.*

Proof.
1. There is at least one pushy flug; call it "*P*".   Th SF3, Def "pushy"
2. Pretend there is another pushy flug "*Q*".   RAA hypothesis
3. *P* scorples *Q*.   1, Def "pushy"
4. *Q* does not scorple *P*.   Th SF1
5. But *Q* must scorple *P*.   2, Def "pushy"
6. Contradiction.   4 and 5
7. Therefore *P* is the only pushy flug.   2–6, logic

Figure 140 summarizes what we know to this point. Representing flugs by circles and scorpling by arrows, there are precisely four flugs (Ax SF4); the unique pushy flug (Th SF4) scorples all the others (Def SF1); no flug scorples the pushy flug (Th SF1); and there is precisely one scorpling relationship (one arrow) joining each pair of outer flugs (Cor to Th SF1).

The figure is rich in suggestions for further theorems. For example, each particular assignment of arrows around the periphery (as in Figure 141) seems

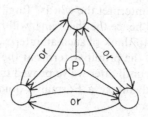

Figure 140

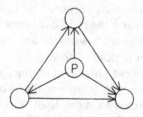

Figure 141

to yield a flug (a "deferential" flug?) that is scorpled by all the others, and does no scorpling itself; and by analogy with the pushy flug we would expect it to be unique. If this could be proven we would be well on the way to individualizing all four flugs.

But enough. I think we've gone far enough for you to see that Formal Axiomatic Systems, murky as they are, still contain things for an intelligence to grasp hold of. It's true that we don't know what a flug is, or what it means for one to scorple another, or what the axioms mean, or what our very own theorems mean. But these apparent handicaps notwithstanding, we have been able to conduct a recognizably mathematical investigation. We have *proven* those meaningless theorems!

How to Not Let the Pictures Bother You

So what do you think?

"What good are meaningless theorems?"

Well, deriving them can be fun. But I guess you're asking for something more practical. That will take a bit longer to explain.

In a Formal Axiomatic System the primitive terms aren't given any meaning. That seems like a weakness. Actually it's a strength, because it frees us to *give* them meanings—to interpret them—in any way that converts the axioms into true statements. Then the theorems become true statements also, and what was meaningless is transformed into real knowledge.

"You mean The Scorpling Flugs can be made into 'real knowledge'?"

Yes, though not of a very interesting kind. The Scorpling Flugs is too simple; it's on about the same level as The Turtle Club in Chapter 1. But it's certainly possible, yes, to interpret the words "flug" and "scorple" in such a way that the axioms, and hence the theorems, will be true. Later you might want to sit down and try to think of such an interpretation; it's not difficult.

"But why not interpret the primitive terms at the very beginning? What's the point to putting it off until after we've proved the theorems?"

You're asking what advantage a Formal Axiomatic System has over a Material Axiomatic System.

One advantage is that with uninterpreted terms it's not as easy to fool ourselves with invalid proofs. Remember, when you were first learning to prove simple theorems in high school geometry, how hard it was not to use "obviously true" things you hadn't proved yet? Even experienced mathematicians can fall into that error if the primitive terms have been explained. The less our imaginations have to work with, the less likely we are to accidentally accept something we don't logically know.

But the major argument for developing a mathematical system "formally"—with uninterpreted terms—is that Formal Axiomatic Systems can sometimes be interpreted in more than one way. This has happened several times in the history of mathematics. There is, for example, a Formal

Axiomatic System called "group theory" that under one interpretation tells about the solvability of algebraic equations, under a second becomes a method for classifying crystals, under a third describes the behavior of elementary particles, and has a number of other interpretations as well.

What makes multiple interpretations possible is our ability to recognize the same logical structure within different concrete situations. If we have studied an axiomatic system formally the likelihood of our perceiving multiple interpretations is greatly increased, for then we at least have a clear conception of what the logical structure is. If on the other hand our experience with a system has been confined within one particular interpretation we are helpless to reinterpret it unless we first carry through the delicate and exasperating process of disentangling its essentials from a clutter of incidentals.

"Can The Scorpling Flugs be interpreted in more than one way?"

Oh, yes. Once you come up with one interpretation you'll be able to think of several others. But they're all pretty boring, the system is so simple.

Look, what do you say we get back to hyperbolic geometry? The only reason we've been talking about Formal Axiomatic Systems is that we are going to consider it one.

"OK."

During our earlier conversation we had drawn this picture [Figure 142] to illustrate the situation that, according to ~Playfair's Postulate, pertains somewhere in the plane. *EF* is distinct from *CD* and parallel to *AB*. Does it look any less objectionable than before?

"Well . . . it still looks like *EF* is going to hit *AB*."

[Drawing Figure 143] What about now?

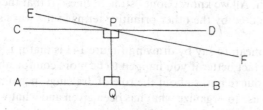

Figure 142

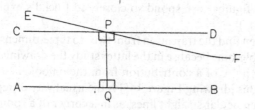

Figure 143

Figure 144

"Now *EF* isn't straight. —Oh, I get it. If hyperbolic geometry is a Formal Axiomatic System then 'straight line' is an undefined term. We don't really know what a 'straight line' will do when we produce it."

We know *EF* won't hit *AB*, so I made the drawing reflect that.

"So in hyperbolic geometry straight lines are curved?!"

If you mean "curved" in the Euclidean sense then they can't be just any old curved lines, for then [drawing Figure 144] it would be possible for two different "straight lines"—that is, Euclidean curved lines—to join the same two points, contrary to Postulate 1. Remember that except for Postulate 5 everything in Euclidean geometry's foundation is also in hyperbolic geometry's foundation.

If on the other hand you mean "curved" in the hyperbolic sense—which is, strictly speaking, the only legitimate sense when we are doing hyperbolic geometry—then hyperbolic straight lines cannot possibly be curved. For what do we mean by "curved line"? I think you'll agree that we mean a "line" (whatever that is) that is *not* a "straight line" (whatever *that* is). Though "line" and "straight line" are undefined primitive terms it is clear, just on logical grounds, that the same thing cannot be both a "straight line" and a "curved line."

No, when I drew Figure 143 I wasn't saying that straight lines bend. I don't know what straight lines do. "Straight line" is just a noise we make, or a phrase we write, to signify one of the types of object hyperbolic geometry is concerned with. All we know about "straight lines" is that they are related to the objects denoted by the other primitive terms in the ways specified in the postulates.

What I did mean to say by drawing Figure 143 is that it is just as good as Figure 142, in fact better if you happen to be more comfortable with it.

Remember our remarks about the role of drawings in geometry? They are visual aids we use to organize what has been given and what we have proved. When we're lucky they even suggest new things capable of proof. But that's all. They are our servants, not our masters, and most certainly not the objects of our study. In plane Euclidean geometry this is easy to lose sight of because the traditional figures correspond so closely to Euclid's explanations of the primitive terms.

Students often find the transition from two- to three-dimensional Euclidean geometry troublesome because in the latter study the drawings are less explicit and so require more of a contribution from the onlooker to be understood. For example, this [drawing Figure 145] is the usual way of representing three mutually perpendicular straight lines, as in a corner of a room. Angle 1 looks right enough, but in order for a beginner to "see" that angles 2 and 3 are also

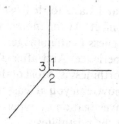

Figure 145

right he must make a conscious effort. For a time he has to keep reminding himself that the old conventions no longer apply, because now the figures are attempting to accommodate an extra dimension.

When a drawing in hyperbolic geometry—which I emphasize is for us a *plane* geometry—attempts to accommodate an extra parallel, the difficulty we experience is analogous. It is impossible to devise a drawing that is completely satisfactory. We have to settle for something like Figure 142 or Figure 143, which are the two types of drawings that are actually used. Much of their initial strangeness stems from our tendency to hang onto the conventions of Euclidean drawings, but once these have been let go we still have to exert ourselves to "see" what they represent. With Figure 142 we have to keep telling ourselves that *EF* won't meet *AB*. With Figure 143 we have to remember that *EF* is just as "straight" as the other lines. Personally I prefer the first kind of figure; there the straight lines all look alike and are "straight" in the everyday sense. And somehow the sentence, "*EF* won't meet *AB*," said while looking at Figure 142, just doesn't catch in my throat the way "*EF* is a straight line" does looking at Figure 143.

"No one has come up with a better way of illustrating hyperbolic geometry?"

You can make pretty good drawings on the surface of a valveless, uncoiled trumpet. The Italian mathematician Eugenio Beltrami (1835–1900) discovered this in 1868. And before the end of the century other ways were devised—practical ones that could be carried out on a flat sheet of paper. But all these amount to *interpretations* of hyperbolic geometry, something I don't want to get into until after the geometry has been developed. If I were to interpret it for you now you'd see our theorems in the context of whatever concrete interpretation I supplied, and the greatest lesson of non-Euclidean geometry would have been lost—that mathematics is, at the core, *completely* abstract, and so about *nothing in particular*.

"So our illustrations will be like Figure 142 or Figure 143?"

Like Figure 142, with the Figure 143 type held in reserve. Yes, our illustrations will be based on the very figure you drew for the first time 'way back on page 161, and which has been central to our discussion ever since. How does it look now?

"Still pretty strange. What I have to do—isn't this right?—is *ignore* its suggestion that *EF* and *AB* meet. At the moment that doesn't seem like it's going to be very easy, but I guess I can manage."

It will get easier with experience. And after a while you'll see how very *limited* the figure's untrustworthiness is. Most of the things it suggests are true, and can be proven. But for now, can you at least see that Figure 142 is a valid tool for the mathematical investigation of hyperbolic geometry?

"I see that it depicts what the postulates say."

That's all it's supposed to do. Good. I think we can begin now.

Exercise

Think of an interpretation for The Scorpling Flugs. That is, think of four objects (the "flugs") and a relation among them ("scorpling") such that Axioms SF1–SF4 (page 164) will be true.

Notes

[1] *they had not seen a proof of* (1) *either*. I'm oversimplifying. They had not seen a *geometric* proof of (1), such as we will examine in Chapter 7. But they could infer the truth of (1) from the representation of Neutral geometry and Postulate 5 within the real number system.

[2] *it is impossible to deduce a contradiction*. We have assumed all long that Euclidean geometry, and so Neutral geometry, is consistent. We will continue to make this assumption. We will examine it in Chapter 7.

[3] *short-circuits the operation of logic itself*. The unsuitability of an inconsistent foundation is obvious to common sense. But in the technical study of logic the issue can be brought into striking focus: if logic is allowed to operate on inconsistent premises, then *every* statement can be proven to be *both* true *and* false and the system evaporates into a cloud of contradictions. See any elementary book on logic.

[4] *clamor*. Letter to F. W. Bessel, January 27, 1829.

[5] *polemic*. Letter to H. K. Schumacher.

[6] *lifetime*. Letter to Farkas Bolyai, March 6, 1832.

[7] *it should not perish with me*. Letter to H. K. Schumacher, May 17, 1831.

[8] *such a remarkable manner*. Letter to Farkas Bolyai, March 6, 1832.

[9] *proof that Playfair's Postulate and Postulate 5 are logically equivalent*. All we have to show (see Chapter 4, exercise 5) is that: Neutral geometry + Postulate 5 $\Rightarrow$ Playfair's Postulate. Our hypothesis, then, is Neutral geometry + Postulate 5. But then we have Theorem 30 ("Straight lines parallel to the same straight line are also parallel to one another"), from which it follows immediately that there cannot be two straight lines that are parallel to the same straight line and pass through the same point.

[10] *Pattern for a Formal Axiomatic System*. From Eves, *A Survey of Geometry* (Allyn and Bacon, 1972), p. 338.

[11] *The Scorpling Flugs*. After Eves, *op. cit.*, p. 340, problem 3.

CHAPTER 6

Hyperbolic Geometry

In elementary books (like this one) the development of hyperbolic geometry is often based on a stronger version of ~ Playfair's Postulate:

Postulate H. If P is any point and AB is any straight line not passing through P (even if produced), then through P there are straight lines YPZ and WPX such that

(1) YPX is not a single straight line,
(2) YPZ and WPX are each parallel to AB, and
(3) no straight line through P entering $\angle\, YPX$ is parallel to AB. (See Figure 146.)

Item (1) is just a way of saying YPZ and WPX are two distinct straight lines.

Postulate H includes ~ Playfair's Postulate, and two extra components besides.

First, it asserts that the more-than-one-parallel phenomenon is universal, occurring at *every* point P in the plane and for *every* straight line AB not passing through it. ~ Playfair's Postulate only guaranteed that the phenomenon occurs at least once, for a particular point and straight line.

Second, the two parallels mentioned in Postulate H are specified as the *lowest* parallels through P in each direction (a straight line through P lower than either would not be parallel to AB by item (3)). The two parallels in ~ Playfair's Postulate were not required to have any additional properties. (Back in Figure 142 we did draw one of the parallels at right angles to the perpendicular from P to AB, but that was on our own initiative. ~ Playfair's Postulate asserts merely that there are two (or more) parallels.)

Both extra components can be proved using ~ Playfair's Postulate and the other hyperbolic axioms (i.e., the axioms of Neutral geometry). Thus Postulate H is really an amalgam of ~ Playfair's Postulate and two *theorems*. However the proofs of those theorems are not easy. By starting with Postulate H the proofs are avoided and the beginner is spared a disheartening first encounter with hyperbolic geometry.

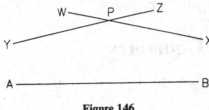

Figure 146

Our path will be to prove the first extra component—for us that will be easy in view of the work we did in Chapter 4—after which I will present a plausibility argument for the second. However, if you skipped the second-last section of Chapter 4 when I advised you (p. 131) that it was optional, you will probably now want to skip the proof of the first extra component, as it depends on those results. (In this case go to the end of the proof of Metatheorem 10, below.) And if you are very impatient to get on with hyperbolic geometry you may want to skip the plausibility argument for the second extra component as well. (In this case start the next section, p. 177.)

Metatheorem 10. *If in the context of Neutral geometry it were known that*

for at least one point and at least one straight line not passing through it (even if produced), there are, through the point, at least two parallels to the straight line,

then it would follow that

for every point and every straight line not passing through it (even if produced), there are, through the point, at least two parallels to the straight line.

(In other words, Neutral geometry + ~ Playfair's Postulate ⇒ the more-than-one-parallel phenomenon is universal.)

Proof. (The first part combines earlier results to show that Neutral geometry + ~ Playfair's Postulate ⇒ no triangle can have an angle-sum of 180°.)

1. In the context of Neutral geometry, Playfair's   p. 172
   Postulate is logically equivalent to Postulate 5.
2. In the context of Neutral geometry, Postulate 5   Cor 3 to Metath 9
   is logically equivalent to the existence of a   (p. 146)
   triangle with angle-sum 180°.
3. Therefore, in the context of Neutral geometry,   1, 2, logic
   Playfair's Postulate is logically equivalent to the
   existence of a triangle with angle-sum 180°.
4. Therefore, in the context of Neutral geometry,   3, logic
   ~ Playfair's Postulate is logically equivalent to
   the *non*existence of a triangle with angle-sum
   180°.

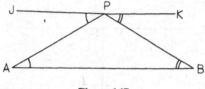

Figure 147

5. But the present context is Neutral geometry,   hypothesis
   and ~ Playfair's Postulate is given.
6. Therefore no triangle has angle-sum 180°.   4, 5
(Now we embark on a proof by contradiction of our desired conclusion.)
7. Pretend there is a point "*P*" and a straight line   RAA hypothesis
   "*AB*" not passing through it (even if produced),
   such that through *P* there is no more than one
   parallel to *AB* (see Figure 147).
8. Draw *PA* and *PB*.   Post 1
9. Through *P* draw "*J*"*P* making ∠*JPA* =   Th 23
   ∠*PAB*.
10. *JP* is parallel to *AB*.   Th 27/28
11. Through *P* draw *P*"*K*" making ∠*KPB* =   Th 23
    ∠*PBA*.
12. *PK* is parallel to *AB*.   Th 27/28
13. But through *P* there is no more than one parallel   7
    to *AB*.
14. Therefore *JPK* is a single straight line.   10, 12, 13
15. ∠*JPA* + ∠*APB* + ∠*KPB* = 180°   Th 13
16. ∠*PAB* + ∠*APB* + ∠*PBA* = 180°   9, 11, 15, C.N. 6 (If
    $a = b, c = d$, and
    $a + e + c = f$ then
    $b + e + d = f$.)
17. Contradiction.   6 and 16
18. Therefore, for every point and every straight   7–17, logic
    line not passing through it (even if produced),
    there are, through the point, at least two paral-
    lels to the straight line.

So much for the first extra component of Postulate H.

The second extra component can be argued pretty convincingly if we introduce the idea of a straight line that moves.

Here once again (Figure 148) is our original drawing of ~ Playfair's Postulate, only now, by what we have just proved, *P* and *AB* are representative of *every* nonincident point and straight line in the plane.

Imagine a straight line, initially coinciding with *PQ*, that pivots around *P*

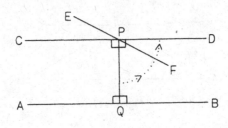

Figure 148

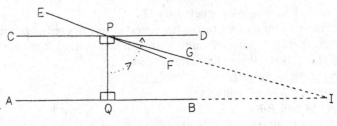

Figure 149

counterclockwise until it reaches the position occupied by *PD*. For a while it will remain nonparallel to *AB*; then it will become parallel to *AB*, as at *PF* and *PD*. Note that once the revolving line starts to occupy positions parallel to *AB*, it will continue to be parallel to *AB* in all subsequent positions during its quarter-turn. For if the revolving line, having assumed a position, say *PF*, parallel to *AB* (Figure 149), were to assume a later position *PG* not parallel to *AB* (so *PG* produced would intersect *AB* produced at *I*), then *PF*, drawn from a point on a triangle (*PQI*) to a point inside, would, if produced, intersect the triangle a second time by Postulate 6 (iii); and the point of intersection could not be on *PQ* or *PI* by Postulate 1; so *PF*, if produced, would intersect *QI*, in contradiction to the fact the *PF* and *AB* are parallel.

Since the moments when the revolving line is *not* parallel to *AB* are all *prior* to the moments when it *is* parallel, there must be a single, sharply-defined moment of *transition*, which can be either a final moment when it is not parallel or a first moment when it is. In other words, during its quarter-turn the straight line must either assume a *last* position not parallel to *AB* or a *first* position parallel to *AB*.

But a last position not parallel to *AB* is impossible, for beyond any given nonparallel position—say *PJ* in Figure 150—one can always find a later nonparallel position: just choose *L* on *QK* produced by Postulates 2 and 10, then draw *PL* by Postulate 1.

Therefore there is a first position parallel to *AB*, which Postulate H calls "*WPX*." The same reasoning on the other side yields Postulate H's "*YPZ*" as

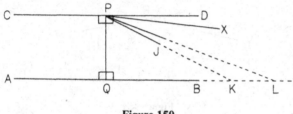

Figure 150

the first position parallel to AB of a straight line revolving clockwise from PQ to PC.

The foregoing is no more than a plausibility argument because it depends on a nongeometric idea, motion. That difficulty can be circumvented, and the argument transformed into a formal proof, but only by technical subtleties I'd rather not get into.

Hyperbolic Geometry (Part 1)

In hyperbolic geometry the primitive terms are the same as in Euclid's geometry: *point, line, straight line, surface*, and *plane surface* (*plane*). We also take over our entire roster of defined terms, to which we will make a few additions as we go along. The axioms are Common Notions 1–6, Postulates 1–4 and 6–10 (all as in Euclidean geometry) and Postulate H (p. 173). As we consequently have all the theorems of Neutral geometry—that is, the "Theorems Without Postulate 5" on pages 98–99, along with Theorems A, B, and C of Chapter 4 (don't worry if you skipped them)—we begin our study of hyperbolic geometry at the somewhat advanced point at which it diverges from Euclidean geometry.

In this section we will concentrate on hyperbolic parallels. Then, in the next section, we will try to reconcile what we have seen with "common sense." We will resume the formal development of hyperbolic geometry in the subsequent section.

In what follows we are concerned exclusively with the points and straight lines of a single plane.

Theorem H1. *In the situation described in Postulate H, every straight line through P that enters* $\angle ZPX$ *is parallel to* AB. (See Figure 151.)

Proof.
1. Let "PC" enter $\angle ZPX$. hypothesis
2. Pretend that PC is not parallel to AB (Figure 152). RAA hypothesis
3. Then PC, if produced sufficiently, will meet AB, or AB produced, at a point "D". Post 2, 2, Def "parallel"

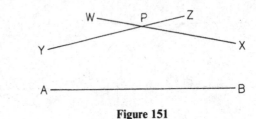

Figure 151

Figure 152

4. Draw $P"Q"$ perpendicular to AB (or AB produced).	Th 12 (Post 2)
5. PX, if produced sufficiently, will meet QD.	Post 2, Post 6 (iii), Post 1
6. But PX, no matter how much it is produced, will never meet QD.	Post H (2), Def "parallel"
7. Contradiction.	5 and 6
8. Therefore PC is parallel to AB.	2–7, logic

Thus the straight lines through P fall into two categories. One contains the infinitely many straight lines entering $\angle YPX$; these, if produced, will intersect AB (or AB produced). The other consists of YPZ, WPX, and the infinitely many straight lines entering $\angle ZPX$; these, no matter how much they are produced, will never intersect AB (or AB produced)—they are parallel to AB. Within the second category YPZ and WPX have somewhat special status as they delineate the division.

Definition H1.[1] In the situation described in Postulate H, the straight lines YPZ and WPX are called the *asymptotic parallels* (or *a-parallels*) *through P to AB*, and the straight lines through P entering $\angle ZPX$ are called the *divergent parallels* (or *d-parallels*) *through P to AB*. (See Figure 153.)

The adjectives "asymptotic" and "divergent" will be explained later on. In the meantime all you need to remember is that the *a*-parallels are the two parallels closest to AB, and that the *d*-parallels are all the others.

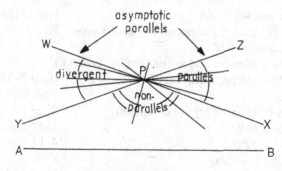

Figure 153

In Figure 152 angles YPQ and XPQ appear to be equal and acute. The next theorem shows that in this respect, at least, our drawings can be trusted.

Theorem H2. *The asymptotic parallels through a point to a straight line make equal, acute angles with the perpendicular from the point to the straight line.*

Proof.

1. Let "P" be a point and "AB" a straight line not   hypothesis
   passing through P (even if produced), and let
   "Y"P and P"X" be the a-parallels through P to
   AB.
2. Draw P"Q" perpendicular to AB (or AB pro-   Th 12 (Post 2)
   duced). (See Figure 154.)

(We will show first that $\angle YPQ = \angle XPQ$.)

3. Pretend $\angle YPQ \neq \angle XPQ$, say $\angle YPQ >$ RAA hypothesis
   $\angle XPQ$.

(The argument would be similar if $\angle XPQ$ were the larger.)

4. Draw P"C" making $\angle CPQ = \angle XPQ$.   Th 23
5. $\angle YPQ > \angle CPQ$ 3, 4, C.N. 6 (If $a > b$
   and $c = b$ then $a > c$.)
6. PC enters $\angle YPQ$.   5

(Step 6 is the crucial step. YP and PX are the a-parallels through P to AB by step 1, and now by step 6 we know PC enters $\angle YPX$. Therefore)

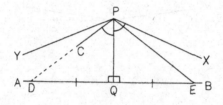

Figure 154

7. *PC* is not parallel to *AB*.   Post H (3)
8. Produce *PC* until it intersects *AB*, or *AB* pro-   Post 2
   duced, at a point "*D*".

(Now we have a triangle, and are on familiar ground.)

9. From *QB* (or *QB* produced) cut off *Q"E"* = *DQ*.   (Post 2) Th 3
10. Draw *PE*.   Post 1
11. Triangles *PDQ* and *PEQ* are congruent.   SAS
12. ∠*CPQ* = ∠*EPQ*   Def "congruent"
13. ∠*XPQ* = ∠*EPQ*   4, 12, C.N. 1
14. But ∠*XPQ* > ∠*EPQ*.   C.N. 5
15. Contradiction.   13 and 14
16. Therefore ∠*YPQ* = ∠*XPQ*.   3–15, logic

(Now we will prove that the equal angles are neither right nor obtuse.)

17. Pretend the equal angles *YPQ* and *XPQ* are   RAA hypothesis
    right (see Figure 155).
18. Then *YPX* is a single straight line.   Th 14

(This is our first use of Theorem 14. Refer to its statement on p. 70. *PQ* is the "straight line", *P* is the "point on it", and *YP* and *PX* are the "two straight lines not lying on the same side" making the adjacent angles, ∠*YPQ* and ∠*XPQ*, add up to 180°.)

19. But *YPX* is not a single straight line.   Post H (1)
20. Contradiction.   18 and 19
21. Therefore angles *YPQ* and *XPQ* are not right.   17–20, logic
22. Now pretend the equal angles *YPQ* and *XPQ*   RAA hypothesis
    are obtuse (see Figure 156).

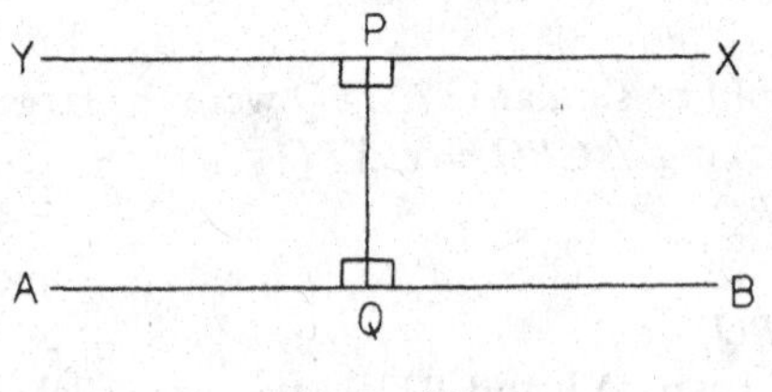

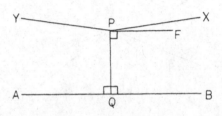

Figure 155

Figure 156

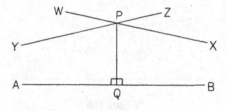

Figure 157

23. Through *P* draw *PF* making a right angle with   Th 11
    PQ.
24. ∠*XPQ* > ∠*FPQ*   22, 23, Def "obtuse"
25. *PF* enters ∠*XPQ*.   24
26. Therefore *PF* is not parallel to *AB*.   Post H (3)
27. But *PF* is parallel to *AB*.   Th 27/28
(The strangeness of hyperbolic parallels notwithstanding, we still have
Theorem 27/28, because it is part of Neutral geometry. It's the *converse*—
Theorem 29—we no longer have.)
28. Contradiction.   26 and 27
29. Therefore angles *YPQ* and *XPQ* are not obtuse.   22–28, logic
30. Therefore the equal angles *YPQ* and *XPQ* are   21, 29, C.N. 6 (If *a* =
    acute.   *b*, then either *a* = *c*
                                                         and *b* = *c*, or *a* > *c*
                                                         and *b* > *c*, or *a* < *c*
                                                         and *b* < *c*.)

We can use Theorem H2 to give individual names to the two *a*-parallels
through *P* to *AB*. We will call *WPX* in Figure 157 the "right" *a*-parallel
through *P* to *AB*, because the acute angle it makes with *PQ* is on the *right* side
of *PQ*; *YPZ* will be the "left" *a*-parallel through *P* to *AB* because its acute
angle is on the left. This terminology is informal, and ambiguous—if you were
to turn the book upside-down, ∠*XPQ* would be on your left and ∠*YPQ* on
your right; but it is also handy, so we will agree, when using it, to look at our
drawings rightside-up.

We will also think of each *a*-parallel as "pointing" in a particular direction,
which we will call its "direction of parallelism." For *WPX* the direction of
parallelism is from *W* to *X*, for *YPZ* it is from *Z* to *Y*.

The next three theorems can be difficult because they say things you may
not realize we don't know yet. You can skip the actual proofs, if you want, but
reading at least the surrounding discussions of what the theorems say, why
they need to be proved, how they will be proved, and their consequences will
shed a lot of light on the nature of hyperbolic parallels.

Given a point *P* and a straight line *AB* not passing through *P* (even if

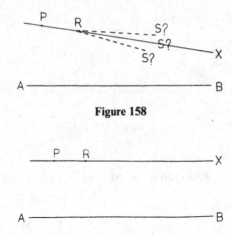

Figure 158

Figure 159. The situation in Euclidean geometry.

produced), let us draw, through P, only the right a-parallel to AB—PX in Figure 158—and consider a point R on PX distinct from P. Then AB doesn't pass through R, either, so Postulate H can be applied to R (and AB), telling us there is a straight line $R"S"$ which is the right a-parallel to AB *through R*. (Postulate H also gives us the left a-parallel to AB through R, but we are not concerned with it.) The question is, what is the relationship between RS and PX? Do they run along together? If not, which one is higher?

You may find my questions strange. You are used to Euclidean geometry where the notion of parallelism is undivided—since there is only one parallel to a straight line through a point, there is only one kind of parallelism to consider. If PX in Figure 159 were the unique Euclidean parallel to AB through P, then it would immediately follow that it is the unique parallel to AB through R as well.

But in hyperbolic geometry we draw a distinction between a-parallelism and d-parallelism, and that distinction is made in reference to *specific* points—note the recurrent phrase "through P" in Definition H1. It is not permissible, at least as far as we know at the moment, to speak of a straight line as being simply "a-parallel" or "d-parallel" to another straight line, without referring to a particular point, because whether or not the straight line *is* a-parallel or d-parallel to the other seems to depend, in Definition H1, on what point we have in mind. It is altogether conceivable that the straight line be a-parallel through one point but d-parallel through some other point.

Back in Figure 158, we know two things about PX: it is parallel to AB, and it is the parallel closest to AB, through P, in that direction (in our drawing, to the right). Its being parallel to AB is a property of the straight line PX *as a whole*—it is parallel to AB when we focus our attention on R just as surely as it is parallel to AB when we focus our attention on P—so there's no difficulty in shifting our focus to R and saying RX is parallel to AB. But there *is* difficulty in shifting the *asymptotic nature* of *PX*'s parallelism at *P* over to *R*. Admittedly it is somewhat natural to suspect that since *PX* is the closest parallel to *AB* through *P*, it is probably the closest parallel to *AB* through *R*, too; but we must guard against assuming this before it has been proved, if indeed a proof is possible. It may be that over at *R*, *PX* is only one of the infinitely many *d*-parallels, and that the closest parallel to *AB*, *RS*, falls below *PX*. (The third possibility illustrated in Figure 158—*RS* above *PX*—is not a real possibility, for by Postulate H (3) no straight line through *R* below the right *a*-parallel *RS* is parallel to *AB*.)

It turns out that our untutored assumption that *RS* runs along *PX* is true after all, as we will shortly prove. Of course everything we've said about right *a*-parallels will also hold for left *a*-parallels, so the theorem leaves the direction unspecified.

Theorem H3. *If a straight line is the asymptotic parallel through a given point, in a given direction, to a given straight line, then it is, through each of its points, the asymptotic parallel in the given direction to the given straight line.*

In Figure 160 I have taken the given direction to be toward the right. *PX* is the right *a*-parallel to *AB* through *P*, and what the theorem says is that *PX* (or *RX*) is the right *a*-parallel to *AB* through *R* as well. Before setting out to prove this, we should think about how we are going to.

To say that *PX* is the right *a*-parallel to *AB through R* means that

(1) *PX* is parallel to *AB* (which we already know), and
(2) *PX* is the closest parallel to *AB*, through *R*, toward the right.

(2) is all we have to prove. Let's restate it more precisely.

Postulate *H* tells us that through *R* there are both a left *a*-parallel and a right *a*-parallel to *AB*. Imagine the left one drawn (*VR* in Figure 161).

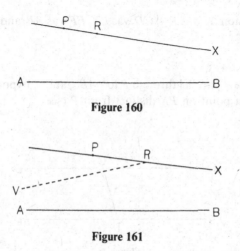

Figure 160

Figure 161

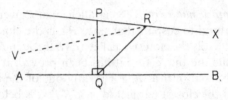

Figure 162

Then what we mean by (2)—I'm getting this from Postulate H—is

(2') no straight line through R entering ∠ VRX is parallel to AB.

This is what we have to prove.

To do so we will draw PQ perpendicular to AB, then RQ (see Figure 162). RQ cuts ∠ VRX into two parts. No straight line through R can be parallel to AB if it enters the left part, ∠ VRQ, because RQ intersects AB and we know VR is the left a-parallel. Therefore, to prove (2') we only have to show

(2") no straight line through R entering ∠ XRQ is parallel to AB.

So the whole proof boils down to proving (2"). And as (2") itself makes no mention of VR, we won't even have to draw VR. Instead we can refer back to this discussion, which for future use we will summarize in the following general principle.

Lemma. *Given a straight line (GI) falling on parallel straight lines (CD and EF), to show that CD is the right (respectively, left) asymptotic parallel through G to EF, it suffices to show that*

(∗) *no straight line through G entering* ∠ DGI *(respectively,* ∠ CGI*) is parallel to EF. (See Figure 163.)*

In our discussion *GI* was *RQ*, *CD* was *PX*, *EF* was *AB*, and the direction was to the right.

Proof of Theorem H3.

1. Let "*PX*" be *a*-parallel through *P* to "*AB*", and   hypothesis
   let "*R*" be a point on *PX* distinct from *P* (see
   Figure 164).

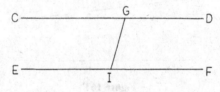

Figure 163

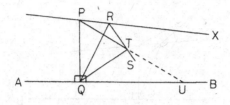

Figure 164

(*R* can be either in the direction of parallelism from *P*—in our drawing, to the right of *P*—or in the other direction from *P*, so there are two cases to consider.)

Case 1. *R* is in the direction of parallelism from *P*.

2. Draw *P*"*Q*" perpendicular to *AB* (produced if necessary), and draw *RQ*.	Th 12 (Post 2), Post 1
3. The straight lines on which *RQ* falls—*PX* and *AB*—are parallel.	1, Post H (2)
4. To show, then, that *PX* is *a*-parallel to *AB* through *R* in the given direction, it suffices to prove (∗) no straight line through *R* entering ∠*XRQ* is parallel to *AB*.	Lemma p. 184
5. Pretend that through *R*, entering ∠*XRQ*, there is a straight line *R*"*S*" which is parallel to *AB*.	RAA hypothesis

(Because of step 4 our attention has shifted to proving (∗), whose form makes it a natural for proof by contradiction.)

6. Choosing a point "*T*" between *R* and *S*.	Post 10 (ii)
7. Draw *PT*.	Post 1
8. *PT* is not parallel to *AB*.	1, Post H (3)
9. Produce *PT* until it intersects *AB*, or *AB* produced, at a point "*U*".	Post 2
10. Draw *TQ*.	Post 1
11. *RS*, if produced, will intersect △*TQU* one more time, at a point "*V*".	Post 2, Post 6 (iii)
12. *V* is not on *TQ* or *TU*.	Post 1
13. Therefore *V* is on *QU* (and so on *AB*).	11, 12
14. Therefore *RS* is not parallel to *AB*.	11, 13
15. Contradiction.	5 and 14
16. Therefore no straight line through *R* entering ∠*XRQ* is parallel to *AB*.	5–15, logic
17. Therefore *PX* is *a*-parallel to *AB* through *R* in the given direction.	4, 16

Case 2. *R* is in the other direction from *P*. (I'll leave the proof of Case 2 as an exercise. It is like the proof of Case 1, except that *T* is taken on *SR* produced. See Figure 165.)

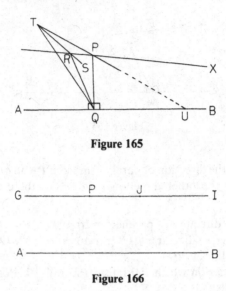

Figure 165

Figure 166

Corollary. *If a straight line is a divergent parallel through a given point to a given straight line, then it is, through* each *of its points, a divergent parallel to the given straight line.*

Proof. Say "*GI*" is *d*-parallel through "*P*" to "*AB*" and let "*J*" be any point on *GI* other than *P* (see Figure 166). By Theorem H1, *GI*, considered as a straight line through *P*, is parallel to *AB*. But whether one straight line is parallel to another is independent of what point on it we refer to, so *GI* is still parallel to *AB* when we transfer our attention to *J*.

All we have to show, then, is that through *J*, *GI* is *d*-parallel rather than *a*-parallel. This is where the theorem comes in: it tells us that if *GI* were *a*-parallel to *AB* through *J*, it would also be *a*-parallel to *AB* through *P*, which we know isn't so.

Theorem H3 frees us from having to mention a particular point when we speak of one straight line being *a*-parallel to another, as the straight line is *a*-parallel to the other through *any* of its points. Similarly, the Corollary frees us from having to mention a particular point when we speak of one straight line being *d*-parallel to another. A straight line is either *a*-parallel or *d*-parallel *as a whole*. Thus Theorem H3 and its Corollary will underlie our every discussion of hyperbolic parallels, though we will rarely refer to them explicitly.

Another question: If one straight line is *a*-parallel to a second, is the second also *a*-parallel to the first? Again our Euclidean experience can cause us to unconsciously assume the answer is yes, when for all we have proved it may be no.

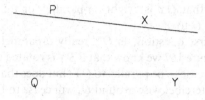

Figure 167

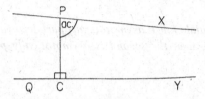

Figure 168

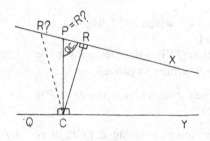

Figure 169

Say, in Figure 167, that PX is right a-parallel to QY. Then PX is, in particular, parallel to QY, which means that PX and QY never meet, no matter how much they are produced—so QY is parallel to PX; that much is simple enough. But *how* is QY parallel to PX? Is it really a-parallel, or could it be d-parallel? And if it *is* a-parallel, in which direction?

We can answer the second question right away. If QY *were* a-parallel to PX it would be the right a-parallel, i.e., the direction of parallelism would be from Q to Y. Here's why (see Figure 168). PX was given to be right a-parallel to QY, so we know the acute angle PX makes with the perpendicular PC is on the right side of PC. Draw CR perpendicular to PX (see Figure 169). If R were to the left of P, the acute angle XPC would be greater than the right angle XRC by Theorem 16—contradiction; and if R were to coincide with P, the acute angle XPC would be equal to the right angle XRC by Common Notion 4—again, a contradiction. So R must be to the right of P, making $\angle YCR$ acute by Common Notion 5. Therefore the acute angle QY would make with the perpendicular CR would be on the *right* side of CR, which is precisely what we mean when we say that QY is "right" a-parallel, or that the "direction of parallelism" is from Q to Y.

This leaves the first question. Is QY really a-parallel, or could it be d-parallel? Back in Figure 167 we know that if PX is rotated clockwise around P even slightly, it will cease to be parallel to QY; but does that mean QY, as soon as it is rotated counterclockwise around Q, will cease to be parallel to PX?

As a matter of fact, it does. (Naiveté wins again.) But surprisingly the proof is quite complex. Due to Lobachevsky, it hasn't been improved on in a hundred and fifty years.

Theorem H4. *If a straight line is asymptotically parallel, in a given direction, to a second straight line, then the second is asymptotically parallel in the same direction to the first.*

Proof.
1. Let "PX" be a-parallel to "QY".   hypothesis
(In the drawing the direction of parallelism is to the right. See Figure 170.)
2. Draw P"C" perpendicular to QY (produced   Th 12 (Post 2)
   if necessary).
3. The straight lines on which PC falls—PX and   1, Post H (2)
   QY—are parallel.
4. To show, then, that QY is a-parallel to PX in the   Lemma, p. 184
   given direction, it suffices to prove

   (∗) no straight line through C entering $\angle YCP$ is
   parallel to PX.

5. Pretend that through C, entering $\angle YCP$, there   RAA hypothesis
   is a straight line CD which is parallel to PX.
(Just as in the proof of Theorem H3, our attention shifts to proving (∗). We will get our contradiction by showing that CD, if produced, will intersect PX (or PX produced). This proof is remarkable in that it gives explicit directions for locating, with compass and straightedge, the point of intersection. To make the main argument easier to follow I have left the verification—requiring only theorems from Neutral geometry—of a few fine points as exercises.)
6. Draw P"E" perpendicular to CD (produced if   Th 12 (Post 2)
   necessary). (See Figure 171).
7. E is on the same side of C as D.   exercise

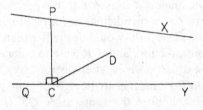

Figure 170

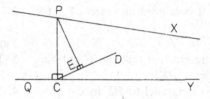

Figure 171

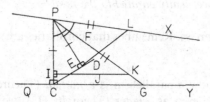

Figure 172

(If—Case 1—*E* should happen to be on the other side of *PX*, then *CD* intersects *PX* (or *PX* produced) by Postulate 7 (ii) and we have our contradiction; so we will assume—Case 2—that *C* and *E* are on the same side of *PX*, as drawn.)

8. *PC* > *PE*	exercise
9. ∠*XPE* < ∠*XPC*	7, hypothesis of Case 2, C.N. 5
10. Through *P* draw *P"F"* making ∠*FPC* = ∠*XPE*.	Th 23
11. ∠*FPC* < ∠*XPC*	9, 10, C.N. 6 (If $a < b$ and $c = a$ then $c < b$.)
12. *PF* enters ∠*XPC*, as drawn in Figure 172.	11
13. *PF* is not parallel to *QY*.	1, Post H (3)
14. Produce *PF* until it intersects *QY*, or *QY* produced, at a point "*G*".	Post 2
15. From *PC* cut off *P"I"* = *PE*.	8, Th 3
16. From *I* draw *I"J"* at right angles to *PC*.	Th 11
17. *IJ* produced meets *PG* at a point "*K*".	exercise
18. From *PX*, or *PX* produced, cut off *P"L"* = *PK*.	(Post 2) Th 3
19. Draw *EL*.	Post 1
20. Triangles *PIK* and *PEL* are congruent.	15, 10, 18, SAS
21. ∠*PEL* = 90°	Def "congruent"
22. But ∠*PED* = 90°.	6
23. Therefore ∠*PEL* = ∠*PED*.	21, 22, C.N. 1
24. Therefore *L* is on *CD* produced.	exercise

(Intuitively, step 24 is obvious. It is an exercise only because its formal verification involves a few steps.)

25. Therefore *CD*, if produced, intersects *PX* (or   24
    PX produced).
26. Contradiction.   5 and 25
27. Therefore no straight line through *C* entering   5–26, logic
    ∠ *YCP* is parallel to *PX*.
28. Therefore *QY* is *a*-parallel to *PX* in the given   4, 27
    direction.

Corollary. *If a straight line is divergently parallel to a second straight line, then the second is also divergently parallel to the first.*

(The proof is an exercise. Remember that corollaries are supposed to be easy to prove!)

Now that we know *a*-parallelism is a *mutual* (Theorem H4) relation of straight lines considered as *wholes* (Theorem H3), we can introduce the following useful convention into our drawings. Whenever two straight lines are *a*-parallel we will enclose them with a brace placed in the direction of parallelism. *D*-parallels and nonparallels we will leave unmarked.

The gain in clarity and expressiveness this unobtrusive device will bring to our drawings is considerable. The braces constantly remind us that, appearances notwithstanding, asymptotic parallels will never meet. And though we make no special indication of other kinds of straight lines, often we will be able to identify a pair as being either *d*-parallel or not parallel from their positions relative to known *a*-parallels. Figure 173, for example, contains the same information as the more complicated Figure 153.

In Theorems H3 and H4 we saw how two properties of Euclidean parallels carry over to hyperbolic *a*-parallels. The next theorem is concerned with the hyperbolic analog of another Euclidean property: "Straight lines parallel to the same straight line are also parallel to one another." This is Euclid's Theorem 30, logically equivalent to Postulate 5 (it is replacement postulate 3A on p. 128) and so false in the present context, as we can see directly from Figure 173: any two straight lines through *P* that do not enter ∠ *YPX* are both parallel to *AB*, but certainly not parallel to one another. No hyperbolic theorem is possible even if we restrict the straight lines to be *a*-parallels: *YZ* and *WX* are *a*-parallel to *AB*, but they intersect at *P*. If however we insist that,

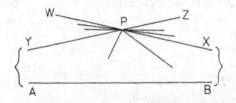

Figure 173

unlike *YZ* and *WX* which are *a*-parallel to *AB* in opposite directions, the straight lines be *a*-parallel to the third straight line in the *same* direction, a theorem is finally possible. Its proof is due to Gauss.

Theorem H5. *If two straight lines are asymptotically parallel, in the same direction, to the same straight line, then they are also asymptotically parallel, in that direction, to one another.*

Proof.

1. Let "*PX*" and "*QY*" be *a*-parallel in the same   hypothesis
direction to "*RZ*".

(This means we can put two braces in our drawing, one enclosing *PX* and *RZ*, the other enclosing *QY* and *RZ*. What we have to prove is that we are entitled to add a third brace enclosing *PX* and *QY*. There are two cases to consider.— see Figure 174—according as *RZ* is or is not between *PX* and *QY*. I will prove one of them, and leave the other for you to try, if you'd like—it involves the same ideas.)

Case 1. *RZ* is not between *PX* and *QY*; say, to be specific, that *QY* is between *PX* and *RZ* (see Figure 175).

(The proof would be the same if *PX* were the one in the middle. Our first job—steps 2–9—is to show, *without* using the forbidden Theorem 30, that *PX* and *QY* are parallel; then we can throw a third straight line across them and use the Lemma on p. 184 to show that they are in fact *a*-parallel in the given direction.)

2. Pretend *PX* and *QY* are not parallel.   RAA hypothesis

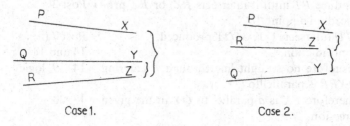

Case 1.   Case 2.

Figure 174

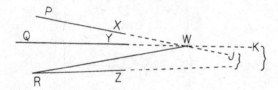

Figure 175

3. Then PX and QY, if produced sufficiently, meet at a point "W". — Def "parallel," Post 2

4. Produce PX beyond W to "J," QY across PJ to "K," and draw RW. — Post 2, Post 1

5. Since QYK is a-parallel in the given direction to RZ, — 1

6. QYK is, in particular, the a-parallel in the given direction to PZ *through* W. — Th H3

7. Therefore PXJ is not parallel to RZ. — Post H (3)

8. Contradiction. — 1 and 7

9. Therefore PX and QY are parallel. — 2–8, logic

10. Draw PR (see Figure 176). — Post 1

11. PR intersects QY, or QY produced, at a point "S". — Post 7 (ii) (Post 2)

12. To show that PX and QY are a-parallel *to one another* in the given direction, it suffices to prove that PX is a-parallel *to* QY in the given direction. — Th H4

13. To show that PX is a-parallel to QY in the given direction, it suffices to prove — 9, Lemma p. 184

   (∗) no straight line through P entering $\angle XPR$ is parallel to QY.

14. Pretend that through P, entering $\angle XPR$, there is a straight line P"T" which is parallel to QY. — RAA hypothesis

15. PX is a-parallel in the given direction to RZ. — 1

16. Therefore PT is not parallel to RZ. — Post H (3)

17. Produce PT until it intersects RZ, or RZ produced, in a point "U". — Post 2

18. PTU intersects QY, or QY produced. — Post 7 (ii)

19. Contradiction. — 14 and 18

20. Therefore no straight line through P entering $\angle XPR$ is parallel to QY. — 14–19, logic

21. Therefore PX is a-parallel to QY in the given direction. — 13, 20

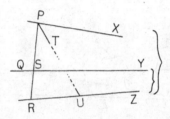

Figure 176

The last three theorems have said things we more or less expected to be true, so you may have found their verification tedious—or even demoralizing, if you had assumed them unconsciously and found yourself having to work hard to gain a position you had thought you already occupied. But from now on things will get better.

Our drawings suggest that a-parallels get closer together in the direction of parallelism. Figure 176, for example, shows PX and RZ approaching each other toward the right. We can prove that this is really happening—here is another respect in which our drawings can be trusted. The proof is easy; it takes advantage of Theorem A, which we can use because it is part of Neutral geometry. If, back in Chapter 4, you skipped Theorem A, you should turn to page 133 and read it now. You needn't read any of the surrounding material, just Theorem A and what there is of its proof.

Theorem H6. *Asymptotic parallels approach each other in the direction of parallelism.*

(To say that two a-parallels "approach" each other in the direction of parallelism means that the perpendiculars drawn from either one to the other get shorter in that direction. In Figure 177, D is any point on PX, C is any point in the direction of parallelism from D, and perpendiculars DA and CB have been drawn. In Figure 178, D and C have been taken on QY instead, and perpendiculars drawn to PX. What we have to prove is that, in either case, CB really is shorter than DA. The proofs of the two cases are similar, however, so it will be enough to prove that CB is shorter in Figure 177.)

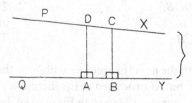

Figure 177

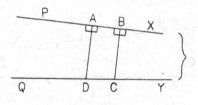

Figure 178

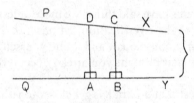

Figure 179

Proof.
1. Let "*PX*" and "*QY*" be *a*-parallels, "*D*" any   hypothesis
   point on *PX*, "*C*" any point on *PX* in the direction of parallelism from *D*, and *D*"*A*" and *C*"*B*"
   perpendiculars to *QY* (see Figure 179).
   (To show: *DA* > *CB*.)
2. ∠*CDA* < 90°   Th H2

(Theorem H2 actually refers to the *two* *a*-parallels through *D* to *QY*, saying
the angles they make with *DA* are equal and acute. When, as now, only one of
the *a*-parallels is drawn, it may seem that the theorem doesn't apply, because
we have only one of the angles. But Theorem H2 still tells us *something*: that
angle is acute. Note, by the way, that we are implicitly using Theorem H3 also:
We know *PX* is *a*-parallel to *QY* *through D* because it is *a*-parallel to *QY*
through *any* of its points. The same reasoning gives us the next step.)

3. ∠*XCB* < 90°   Th H2
4. But ∠*DCB* + ∠*XCB* = 180°,   Th 13
5. so ∠*DCB* > 90°,   3, 4, C.N. 6 (If *a* < *b*
                                     and *c* + *a* = 2*b* then
                                     c > *b*.)
6. and ∠*CDA* < ∠*DCB*.   2, 5, C.N. 6 (If *a* < *b*
                                     and *c* > *b* then *a* < *c*.)
7. Therefore *DA* > *CB*.   Th A

We can say more. In the next theorem we will show that *a*-parallels not only
approach each other, but, if produced in the direction of parallelism, become
arbitrarily close to one another! That is, in Figure 180, I can make the length of
CB as close to 0 as I like simply by placing point *C* far enough to the right. (Of
course, I can't ever make it *equal* to 0, because *PX* and *QY* are parallel.)

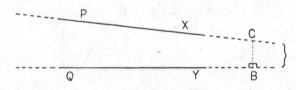

Figure 180

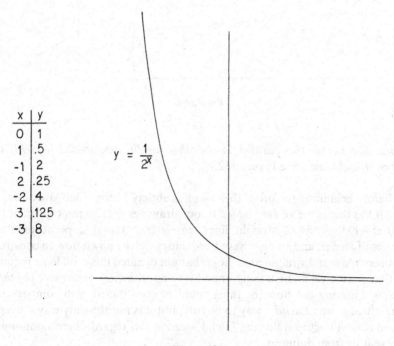

x	y
0	1
1	.5
-1	2
2	.25
-2	4
3	.125
-3	8

$$y = \frac{1}{2^x}$$

Figure 181

This is why asymptotic parallels are called "asymptotic." In algebra an "asymptote" to the graph of an equation is a straight line to which the graph becomes arbitrarily close but never meets. Figure 181 shows the graph of $y = 1/2^x$. Also shown are a few pairs of numbers I used to draw it. (Take my word on them if you find that over the years your algebra has evaporated.) Only part of the graph is shown; it continues to the right, forever, getting ever closer to, but never meeting, the x-axis. We know it will never meet the x-axis because, no matter how huge a number we use for x, $1/2^x$, though tiny, it will never be 0. (Your intuition may cry that the graph and the x-axis meet "at infinity," but that's tantamount to saying they *never* meet. Infinity is not a place; ∞ is not a number.) The relation between PX and QY in Figure 180 is analogous. There is one difference: though the graph of $y = 1/2^x$ gets straighter and straighter toward the right, it never actually becomes straight; but PX and QY are both straight lines throughout.

The analogy extends further. The graph continues forever to the left, also, getting ever more distant from the x-axis; and in Figure 180, I can make the length of CB as large as I like by placing C far enough to the left. In sum, then: for any length whatever—small, large, or in between, so long as it's not 0—I can choose a location for C that will give CB that particular length.

Theorem H7. *Given any pair of asymptotic parallels and any straight line, on each parallel (produced if necessary) there is a point from which the per-*

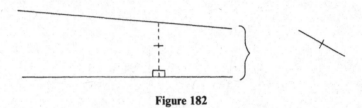

Figure 182

pendicular to the other parallel (or the other parallel produced) is equal to the given straight line. (See Figure 182.)

Before beginning the proof there is one subtlety I should tell you about.

All the things we've ever *added* to our drawings that weren't given at the outset—extensions of straight lines, equilateral triangles, perpendiculars, copies of angles, and so on—have been things we've known how to construct with compass and straightedge. Logic has not dictated this—all logic requires is that, before we name a thing or add it to our drawing, *we be sure the thing exists*. Figuring out how the thing could be constructed with compass and straightedge was *Euclid*'s way to be sure; but it is not the only way. Given a good reason to stop following Euclid's lead in this regard, there's nothing to prevent us from doing so.

We have a good reason. In the upcoming proof we will encounter the situation depicted in Figure 183. We will want to draw the left *a*-parallel through *T* to *QY*. There *is* a compass-and-straightedge method for doing so—discovered by János Bolyai—but proving the method is correct is too difficult for this kind of book. So what will we do? Draw the left *a*-parallel anyway! We are sure it exists, by Postulate H (which we will cite as our reason). We will draw it—*TZ* in Figure 184—to reflect what we know about it: it goes toward

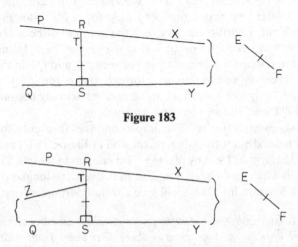

Figure 183

Figure 184

the left, it approaches but never meets QY, and $\angle ZTS$ is acute. Geometric diagrams are only visual summaries of what we know, so it doesn't matter if we've made $\angle ZTS$ a hair too small or too large.

Proof of Theorem H7.

1. Let "PX" and "QY" be the a-parallels, and   hypothesis
   "EF" the straight line.
2. Choose "R" at random on PX, and draw R"S"   Post 10, Th 12
   perpendicular to QY (produced if necessary).   (Post 2)

(There are three cases to consider, according as RS is equal to, greater than, or less than EF. If—Case 1—RS is equal to EF, we're done, so we'll go on to Case 2.)

Case 2. $RS > EF$ (see Figure 183).

3. From RS cut off "T"$S = EF$.   Th 3
4. Draw ZT a-parallel to QY in the direction   Post H
   opposite to the direction in which PX and QY
   are a-parallel.

(We now have the situation of Figure 184. Steps 5–16 will show that ZT is not parallel to PX.)

5. Draw T"W" a-parallel to QY in the same direc-   Post H
   tion in which PX and QY are a-parallel.
6. Then, in the direction in which PX and QY are   Th H5
   a-parallel, TW and PX are a-parallel.
7. Produce ZT to "U" (Figure 185).   Post 2
8. $\angle ZTS < 90°$ Th H2
9. $\angle RTU = \angle ZTS$ Th 15
10. $\angle RTU < 90°$ 8, 9, C.N. 6 (If $a < b$
     and $c = a$ then $c < b$.)
11. $\angle WTS < 90°$ Th H2
12. $\angle RTW + \angle WTS = 180°$ Th 13
13. $\angle RTW > 90°$ 11, 12, C.N. 6 (If
     $a < b$ and $c + a = 2b$
     then $c > b$.)
14. $\angle RTU < \angle RTW$ 10, 13, C.N. 6 (If
     $a < b$ and $c > b$ then
     $a < c$.)
15. Therefore ZTU enters $\angle RTW$, as drawn,   14

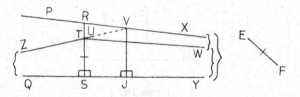

Figure 185

16. so ZTU is not parallel to PX.   6, 15, Post H (3)
17. Produce ZTU until it intersects PX, or PX Post 2
    produced, at a point "V".
18. Draw V"J" perpendicular to QY (produced if   Th 12 (Post 2)
    necessary).

(What we're looking for is a point on PX from which the perpendicular to QY is equal to EF. At this juncture the drawing—the crazy hyperbolic drawing—actually helps. Focus your attention on the portion I've copied in Figure 186. We know that ZV and VX are the left and right a-parallels to QY through V, and that the angles they make with VJ are equal. Note that if we were at V and wanted to find a point *on ZV* from which the perpendicular is equal to EF, we would only have to travel left to T. The symmetric arrangement of ZV and VX then suggests we could find a similar point on VX by simply traveling the same distance to the right!)

19. From VX, or VX produced, cut off V"K" =   Th 3 (Post 2)
    TV (see Figure 187).
20. Draw K"M" perpendicular to QY (produced if   Th 12 (Post 2)
    necessary).
21. Draw TJ and KJ.   Post 1
22. $\angle TVJ = \angle KVJ$ Th H2
23. Triangles TVJ and KVJ are congruent.   SAS
24. $TJ = KJ$ and $\angle VJT = \angle VJK$ Def "congruent"
25. $\angle TJS = \angle KJM$ 18, 24, C.N. 3
26. Triangles TSJ and KMJ are congruent.   AAS
27. $TS = KM$ 26, Def "congruent"
28. $KM = EF$ 3, 27, C.N. 1

(Case 3—$RS < EF$—is similar. I'll leave it as an exercise for you to try, if you'd like. Figure 188 is analogous to Case 2's Figure 185.)

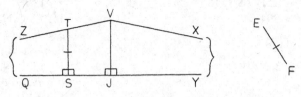

Figure 186

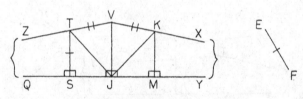

Figure 187

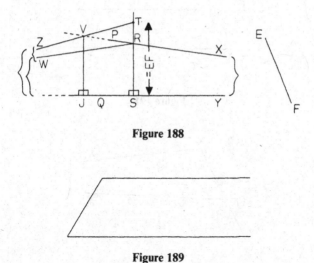

Figure 188

Figure 189

The next theorem is the analog for hyperbolic geometry of Euclid's Theorem 29. Theorem 29 was the first theorem Euclid proved with Postulate 5 (it is in fact logically equivalent to Postulate 5), and concerned what Euclid called "a straight line falling on parallel straight lines," of which Figure 189 is a truncated version. We will begin by giving the corresponding hyperbolic figure a shorter name.

Definition H2. If, from the endpoints of a given finite straight line, and on the same side of it, two straight lines are drawn which are asymptotically parallel to one another in the direction away from the given straight line, the resulting figure is called a *biangle*[2]; the given finite straight line is called its *base*.

Thus $XABY$ (Figure 190) is a biangle. It has only two angles, of course—thus the name—because AX and BY, no matter how much produced, never meet. AB is the base. $WABZ$ in Figure 190 is *not* a biangle, because the direction of parallelism is toward AB, rather than away from AB as it should be.

Producing Euclid's transversal in Figure 189 gives us $\angle 1$ in Figure 191,

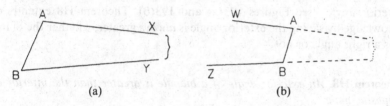

(a)   (b)

Figure 190. (a) A biangle. (b) Not a biangle.

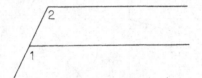

Figure 191

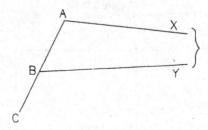

Figure 192

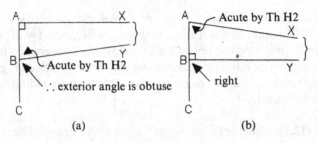

Figure 193

which he called an "exterior" angle. He proved in Theorem 29 that it is equal to the "interior and opposite" angle, $\angle 2$.

Analogously, producing AB in Figure 190 to C (Figure 192), we will call $\angle YBC$ an "exterior" angle, and $\angle XAB$ the corresponding "interior and opposite" angle. (Were we to produce AB beyond A instead, we would call that angle an "exterior" angle, too, in which case the "interior and opposite" angle would be $\angle ABY$.) If either angle of biangle $XABY$ is right we already know, by Theorem H2, that the exterior angle is *greater* than the opposite interior angle—see Figures 193(a) and 193(b). Theorem H8 extends our knowledge by saying the exterior angle is *always* greater, whether the biangle has a right angle or not.

Theorem H8. *An exterior angle of a biangle is greater than the interior and opposite angle.*

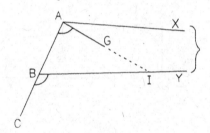

Figure 194

Proof.

1. Let "*XABY*" be a biangle with the base *AB*   hypothesis
   produced to "*C*" (see Figure 194).
2. *AX* and *BY* are *a*-parallel in the direction away   Def "biangle"
   from *AB*.

(We have to show ∠*YBC* is larger than ∠*XAB*. We will do this by showing, first, that ∠*YBC* is not smaller than ∠*XAB* and, second, that ∠*YBC* is not equal to ∠*XAB*, either.)

3. Pretend ∠*YBC* < ∠*XAB*.   RAA hypothesis
4. Through *A* draw *A*"*G*" making ∠*GAB* =   Th 23
   ∠*YBC*.
5. ∠*GAB* < ∠*XAB*   3, 4, C.N. 6 (If *a* < *b*
      and *c* = *a* then *c* < *b*.)
6. *AG* enters ∠*XAB*, as drawn.   5
7. *AG* is not parallel to *BY*.   2, 6, Post H (3)
8. Produce *AG* until it intersects *BY*, or *BY*   Post 2
   produced, at a point "*I*".
9. ∠*YBC* > ∠*GAB*   Th 16 (△*ABI*)
10. Contradiction.   4 and 9
11. Therefore ∠*YBC* ≮ ∠*XAB*.   3–10, logic

(Now we will show that if ∠*YBC* were equal to ∠*XAB* it would be possible to construct a straight line perpendicular to *both AX* and *BY*, in conflict with Theorem H2. This part of the proof presumes that ∠*YBC* is not right. If it were, as in Figure 193(b), the conclusion ∠*YBC* > ∠*XAB* would follow *directly* from Theorem H2.)

12. Pretend ∠*YBC* = ∠*XAB*.   RAA hypothesis
13. Bisect *AB* at "*M*"; draw *M*"*J*" perpendicular   Th 10, Th 12, Post 2,
    to *BY* (produced if necessary); produce *XA* and   Th 3, Post 1
    cut off "*K*"*A* = *BJ*; draw *KM* (see Figure 195).

(Note that we don't know whether *KMJ* is a single straight line or not. Steps 14–21 show that it is.)

14. ∠*MBJ* + ∠*YBC* = ∠*KAM* + ∠*XAB*   Th 13, C.N. 1

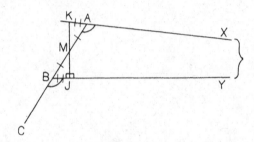

Figure 195

15. $\angle MBJ = \angle KAM$	12, 14, C.N. 3
16. Triangles MBJ and MAK are congruent.	SAS
17. $\angle BMJ = \angle KMA$	Def "congruent"
18. $\angle BMJ + \angle AMJ = \angle KMA + \angle AMJ$	17, C.N. 6 (If $a = b$ then $a + c = b + c$.)
19. But $\angle BMJ + \angle AMJ = 180°$.	Th 13
20. Therefore $\angle KMA + \angle AMJ = 180°$,	18, 19, C.N. 1
21. so KMJ is a single straight line.	Th 14
22. Also $\angle XKJ = 90°$.	16, Def "congruent"
23. But KX is a-parallel to BY through K,	2, Th H3
24. and KJ is perpendicular to BY,	21, 13
25. so $\angle XKJ$ is acute.	Th H2
26. Contradiction.	22 and 25
27. Therefore $\angle YBC \neq \angle XAB$.	12–26, logic
28. Therefore $\angle YBC > \angle XAB$.	11, 27, C.N. 6 (If $a \not< b$ and $a \neq b$ then $a > b$.)

Notice that in our drawings of biangles, it *looks* as if Theorem H8 were true—yet another respect in which our drawings can be trusted.

Unlike Euclid's Theorem 29, Theorem 27/28 is part of Neutral geometry and so is still valid. Its hypothesis is that a straight line falling on two straight lines causes any one of the eight angle relationships (p. 84) to occur; its conclusion, that the two straight lines are parallel. Back in Euclidean geometry that conclusion was perfectly definite. No longer: "All right," we can say, "the two straight lines are parallel. But *how* are they parallel?" Theorem H9 provides the answer.

Theorem H9. *If a straight line falling on two straight lines causes any one of the eight angle relationships to occur, then the two straight lines are divergently parallel.*

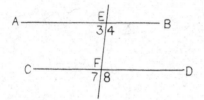

Figure 196

Proof.

1. Let "*AB*" and "*CD*" be the two straight lines   hypothesis
   and "*EF*" the straight line intersecting them—
   at *E* and *F* respectively—that causes any one of
   the eight angle relationships to occur (see Figure
   196).
2. Then *AB*‖*CD*.   Th 27/28
3. Also, ∠7 = ∠3 and ∠8 = ∠4.   1, Lemma p. 84
4. Pretend that *AB* and *CD* are *a*-parallel toward   RAA hypothesis
   the right.
5. Then *BEFD* is a biangle,   Def "biangle"
6. so ∠8 > ∠4.   Th H8
7. Contradiction.   3 and 6
8. Therefore *AB* and *CD* are not *a*-parallel toward   4–7, logic
   the right.
9. Pretend that *AB* and *CD* are *a*-parallel toward   RAA hypothesis
   the left.
10. Then *AEFC* is a biangle,   Def "biangle"
11. so ∠7 > ∠3.   Th H8
12. Contradiction.   3 and 11
13. Therefore *AB* and *CD* are not *a*-parallel toward   9–12, logic
    the left, either.
14. Therefore *AB* and *CD* are *d*-parallel.   2, 8, 13

Theorem H10 (AB). *If an angle and the base of one biangle are equal, respectively, to an angle and the base of another biangle, then the other pair of angles are equal.* ("AB"—for "angle-base"—is the theorem's nickname, not to be confused with any *straight line AB*. See Figure 197).

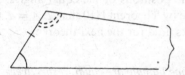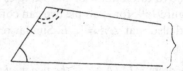

Figure 197

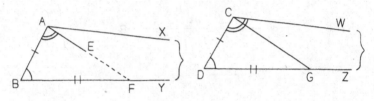

Figure 198

Proof.

1. Let "$XABY$" and "$WCDZ$" be biangles with   hypothesis
   $AB = CD$ and, say, $\angle ABY = \angle CDZ$ (see
   Figure 198).

(To show: $\angle XAB = \angle WCD$.)

2. Pretend $\angle XAB \neq \angle WCD$, say $\angle XAB >$ RAA hypothesis
   $\angle WCD$.

(The argument would be similar if $\angle WCD$ were the larger.)

3. Through A draw $A``E"$ making $\angle EAB =$ Th 23
   $\angle WCD$.

4. $\angle EAB < \angle XAB$ (So AE enters $\angle XAB$, as   2, 3, C.N. 6 (If $a > b$
   drawn.)   and $c = b$ then $c < a$.)

5. AE is not parallel to BY.   4, Post H (3)

6. Produce AE until it intersects BY, or BY Post 2
   produced, at a point "F".

7. From DZ, or DZ produced, cut off $D``G" =$ (Post 2) Th 3, Post 1
   BF, and draw CG.

8. Triangles ABF and CDG are congruent.   SAS

9. $\angle EAB = \angle GCD$ Def "congruent"

10. Therefore $\angle WCD = \angle GCD$.   3, 9, C.N. 1

11. But $\angle WCD > \angle GCD$.   C.N. 5

12. Contradiction.   10 and 11

13. Therefore $\angle XAB = \angle WCD$.   2–12, logic

To make the argument easier to follow, the biangles in Figure 198 were oriented the same way and the equal angles placed in the same corner. It is clear from the proof, however, that Theorem H10 applies regardless of the relative orientation of the biangles or the positions of the equal angles. In Figure 199, for example, we can conclude by Theorem H10 that $\angle A = \angle C$, and also that $\angle A = \angle E$. Similar remarks hold for the next theorem.

Theorem H11 (AA). *If the two angles of one biangle are equal, respectively, to the two angles of another biangle, then the bases are equal.* (See Figure 200.)

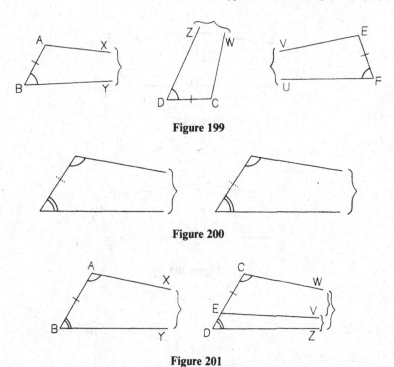

Figure 199

Figure 200

Figure 201

Proof.

1. Let "*XABY*" and "*WCDZ*" be biangles with,   hypothesis
   say, $\angle XAB = \angle WCD$ and $\angle ABY = \angle CDZ$
   (see Figure 201).

(To show: $AB = CD$.)

2. Pretend $AB \neq CD$, say $AB < CD$.   RAA hypothesis

(The argument would be similar if AB were greater than CD.)

3. From CD cut off $C"E" = AB$.   Th 3

4. Through E draw $E"V"$ a-parallel to DZ in the   Post H
   same direction in which CW and DZ are
   a-parallel.

5. Then EV is also a-parallel in that direction   Th H5
   to CW.

6. $WCEV$ is a biangle.   5, Def "biangle"

7. $\angle ABY = \angle CEV$ AB (Th H10)

8. $\angle CDZ = \angle CEV$ 1, 7, C.N. 1

9. But $VEDZ$ is a biangle, too,   4, Def "biangle"

10. making $\angle CEV > \angle CDZ$.   Th H8

11. Contradiction.   8 and 10

12. Therefore $AB = CD$.   2–11, logic

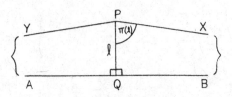

Figure 202

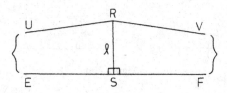

Figure 203

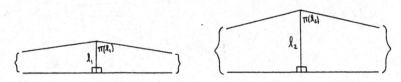

Figure 204

In Theorem H11 we encounter for the first time the curious fact that in hyperbolic geometry a length can be determined by angles alone. We will meet this phenomenon again.

Lobachevsky called $\angle XPQ$ in Figure 202 the "angle of parallelism" and denoted it by the symbol $\pi(l)$—read "pi of l"—where l is the length of PQ. (In this context the Greco-Russian letter "π," corresponding to our letter "p," stands for "parallelism," and has nothing to do with the *number* $\pi = 3.14159\ldots$.) By Theorem H2, $\pi(l) = \angle XPQ = \angle YPQ < 90°$.

The design of the symbol $\pi(l)$ suggests that the indicated angle-size does not depend on the particular point P or straight line AB, but only on the distance l between them. This is true. For if R and EF, in Figure 203, are any other point and straight line separated by the same distance—that is, if the perpendicular RS also has length l—then $\angle VRS = \angle XPQ$ by AB (Theorem H10).

But the angle-sizes will be different if the *distances* are different. Given, in Figure 204, $l_1 \neq l_2$, prove $\pi(l_1) \neq \pi(l_2)$. Pretend, as an RAA hypothesis, $\pi(l_1) = \pi(l_2)$; then $l_1 = l_2$ by AA (Theorem H11), a contradiction.

Thus there is a one-to-one correspondence between the various distances l and the corresponding angle-sizes $\pi(l)$.

Reconciliation With Common Sense

By developing the properties of hyperbolic parallels as much as we have, we have to some degree gotten "used" to their strange behavior, and to a considerable degree come to know the *extent* of their deviation from the path of "common sense." It's time to reactivate our common sense, and see if it's possible to effect some measure of reconciliation between it and the new geometry.

The characteristic hyperbolic situation is shown in Figure 205. *YPZ* and *WPX* are the *a*-parallels through *P* to *AB*, and the perpendicular *CPD* to *PQ* is only one of the infinitely many *d*-parallels through *P* to *AB*. But common sense tells us *CPD* is the *only* parallel through *P* to *AB*. Is it possible to reconcile these two positions?

What if the angle of parallelism $\pi(l)$ were very close to a right angle? So close that, even in the most meticulously drawn figure, *PX* would appear to pass through *D*? Say $\pi(l)$ were 89.9999999995°. Then $\angle DPX$ and $\angle ZPD$ would each be .0000000005°, so $\angle ZPX$ would be .000000001°—a billionth of a degree. An accurate drawing would then look like Figure 206, and one would have to produce *PD* and *PX* an astronomical distance to the right in order for the space between them to become visible.

What I am suggesting is that *hyperbolic geometry might actually be true*, and the reason we have never noticed it is that $\angle ZPX$ is so small. From this perspective Postulate H becomes as intuitively acceptable as the notion that water is in fact composed of invisible molecules, or that the apparently flat surface of a pond is really curved.

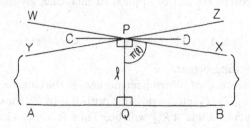

Figure 205

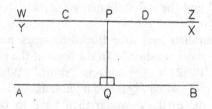

Figure 206

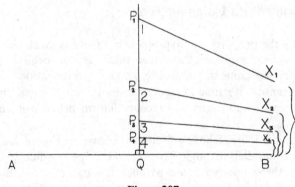

Figure 207

But *can* $\angle ZPX$ be that small? Does hyperbolic geometry allow $\pi(l)$ to be so nearly right? Theorem H8 suggests the answer: Yes, if l is small enough. The plausibility of this can be argued as follows. From point Q in Figure 207 draw a straight line perpendicular to AB and let P_1 be any point on it. Use Theorem 10 over and over to get a series of points P_2, P_3, P_4, etc. such that P_2 is halfway from P_1 to Q, P_3 is halfway from P_2 to Q, P_4 is halfway from P_3 to Q, and so on. Use Postulate H, over and over, to draw right a-parallels $P_1 X_1$, $P_2 X_2$, $P_3 X_3$, etc. to AB. Each of the corresponding angles of parallelism (numbered in the figure) is less than a right angle by Theorem H2. But by Theorem H5, over and over, each of the straight lines $P_1 X_1$, $P_2 X_2$, $P_3 X_3$, etc. is right a-parallel to the straight line immediately below it, so we have a stack of biangles; Theorem H8 can be applied to each one, giving us the series of inequalities

$$\angle 1 < \angle 2, \ \angle 2 < \angle 3, \ \angle 3 < \angle 4, \text{ etc.}$$

The angles are getting bigger.

We have, therefore, the following phenomenon: the closer a point P is taken to AB (think of P_1, P_2, P_3, etc. as possible positions of P), the closer to 90° the corresponding acute angle XPQ will be. Thus it looks like we can make $\angle XPQ = \pi(l)$ as close to 90° as we like simply by taking P close enough to Q. In particular it seems that a point P placed sufficiently near Q would make an angle of parallelism $\pi(l)$ at least as large as the 89.9999999995° value we mentioned before,[3] and the infinitely many parallels through this P would appear to merge as in Figure 206.

So hyperbolic parallels look like Euclidean ones provided that PQ is short. Ah, but how short is "short"? In the scale of the universe two million light-years (about 18.922×10^{18} km) is "short." What if $\angle XPQ$ were 89.9999999995° or more when PQ is two million light-years long?[4] Placing Q on the surface of the earth, P would then have to be further than the Andromeda galaxy (!) before $\angle XPQ$ would be detectably less than a right angle, and any human-scale application of Postulate H would look, if drawn accurately, like Figure 206.

The key to rendering Postulate H imaginable is therefore to suppose that the universe is constructed in such a way that what I've described actually happens. Under this supposition we would think of Postulate H as true—and Euclid's Postulate 5 as false—but in a way that would not contradict human experience.

Common sense would accept the overthrow of Postulate 5 rather calmly, because the truth of Postulate H would be so *abstract*. On the one hand, the accuracy of a figure like Figure 205, that conspicuously displays the hyperbolic behavior of parallels, could be conceived of only abstractly, because the intergalactic distances it would represent so utterly transcend the scale of human activity; on the other hand, the truth of Postulate H in the face of a human-scale figure like Figure 206 would involve the equally abstract notion that the apparently unique parallel through *P* is "really" a bundle of infinitely many parallels contained within an imperceptible angle. Astronomical distances and invisible angles are outside the usual province of common sense, so it has little prejudice concerning them.[5]

Notice how we have used one hyperbolic oddity—Theorem H8—to mitigate the intuitive shock of another—Postulate H. This is no coincidence. Every ostensible paradox in hyperbolic geometry is offset by a complementary one.

Hyperbolic Geometry (Part 2)

In this section we resume the formal development of hyperbolic geometry, which we will carry about as far as we carried our study of Euclid's geometry. We will say a bit more, informally, in the following section.

If you skipped the second-last section of Chapter 4, you should now start with the Definition of "Saccheri quadrilateral" on p. 132, read through the third sentence after the proof of Theorem A's Corollary, then skip to the statement and proof of Theorem B. Theorem A, its Corollary, and Theorem B are all part of Neutral geometry.

Theorem H12. *The base and summit of a Saccheri quadrilateral are divergently parallel, as are the other two sides.*

Proof.

1. Let "*ABCD*" be a Saccheri quadrilateral with   hypothesis
   summit *DC* (see Figure 208).
2. Bisect *DC* at "*M*", *AB* at "*N*", and draw *MN*.   Th 10, Post 1
3. Since ∟*DMN* and ∟*MNB* are both right,   Corollary to Th A
4. ∟*DMN* = ∟*MNB*.   Post 4

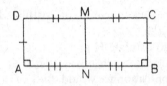

Figure 208

5. Therefore DC and AB are d-parallel.   Th H9 (MN falling
                                                           on DC and AB)
6. Similarly, since $\angle DAB$ and $\angle CBA$ are right,   Def "Saccheri quad."
7. $\angle DAB + \angle CBA = 180°$ C.N. 2
8. so DA and CB are d-parallel, too.   Th H9 (AB falling on
                                                           DA and CB)

We'll use Theorem H12 to verify the diagram in the next proof.

Theorem H13. *The summit angles of every Saccheri quadrilateral are acute.*

Proof.
1. Let "$ABCD$" be a Saccheri quadrilateral with   hypothesis
   summit DC (see Figure 209).
2. Produce DC to "E" and AB to "Z".   Post 2
3. Through D and C draw D"Y" and C"X"   Post H
   a-parallel to ABZ in the direction toward Z.

(The proof will hinge on the fact that DY and CX are under DCE, as drawn.
Our first job, then, is to verify the diagram.)
   4. DCE is d-parallel to ABZ.   Th H12
   5. In particular, then, DCE is d-parallel to ABZ Corollary to Th H3
      through D.
   6. But DY is a-parallel to ABZ through D.   3
   7. Therefore DY enters $\angle EDA$, as drawn.   Def "a-parallel,"
                                                      "d-parallel"
   8. Similarly CX enters $\angle ECB$.   imitate 4–7

(Now we will show that $\angle DCB$ is smaller than $\angle ECB$. As their sum is 180°,
this will mean $\angle DCB$ is acute.)

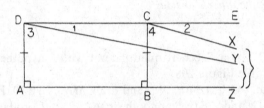

Figure 209

9. CX and DY are a-parallel to one another in the   Th H5
   same direction in which they are a-parallel to
   ABZ.
10. Therefore $XCDY$ is a biangle,   Def "biangle"
11. and $\angle 1 < \angle 2$.   Th H8
12. $YDAZ$ and $XCBZ$ are biangles with   Def "biangle"
13. $\angle DAZ = \angle CBZ$ and $DA = CB$,   Def "Saccheri quad.",
                                                            Post 4
14. so $\angle 3 = \angle 4$.   AB
15. Therefore $\angle CDA < \angle ECB$.   11, 14, C.N. 6 (If
                                                            $a < b$ and $c = d$ then
                                                            $a + c < b + d$.)
16. But $\angle CDA = \angle DCB$,   Corollary to Th A
17. so $\angle DCB < \angle ECB$.   15, 16, C.N. 6 (If
                                                            $a < b$ and $a = c$ then
                                                            $c < b$.)
18. Since $\angle DCB + \angle ECB = 180°$,   Th 13
19. $\angle DCB < 90°$ and   17, 18, C.N. 6 (If
                                                            $a < b$ and $a + b = c$
                                                            then $a < \frac{1}{2}c$.)
20. $\angle CDA < 90°$.   16, 19, C.N. 6 (If
                                                            $a = b$ and $b < c$ then
                                                            $a < c$.)

Theorem H14. *The angle-sum of every triangle is less than* $180°$.

Proof.
1. Let "ABC" be any triangle (see Figure 210).   hypothesis
2. Bisect AB at "D" and AC at "E".   Th 10
3. Draw DE and produce it in both directions.   Post 1, Post 2
4. Draw B"F" and C"G" perpendicular to DE Th 12
   produced.
5. Then $GFBC$ is a Saccheri quadrilateral with   Th B
   summit BC, and $\angle FBC + \angle GCB =$ angle-sum
   of $\triangle ABC$.

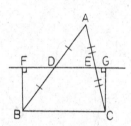

Figure 210

6. Since $\angle FBC < 90°$ and $\angle GCB < 90°$,   Th H13

7. $\angle FBC + \angle GCB < 180°$;   6, C.N. 6 (If $a < b$
                                             and $c < b$ then
                                             $a + c < 2b$.)

8. therefore, the angle-sum of $\triangle ABC < 180°$.   5, 7, C.N. 6 (If $a = b$
                                             and $a < c$ then $b < c$.)

Corollary. *The angle-sum of every quadrilateral is less than* $360°$.

(A "quadrilateral" is any four-sided figure. The easy proof is an exercise.)

After seeing SAS, ASA, AAS, and SSS, many high school students think there must be an AAA theorem as well. Of course there isn't—in Euclidean geometry. Now there is.

Theorem H15 (AAA). *If the three angles of one triangle are equal, respectively, to the three angles of another triangle, then the triangles are congruent.*

Proof.

1. Let "ABC" and "DEF" be triangles with, say,   hypothesis
   $\angle ABC = \angle DEF$, $\angle ACB = \angle DFE$,   and
   $\angle BAC = \angle EDF$ (see Figure 211).

(The idea is to show that one pair of corresponding sides are equal, then use ASA or AAS.)

2. Pretend $AB \neq DE$, say $AB > DE$.   RAA hypothesis

(The argument would be similar if we took DE to be the larger.)

3. From AB cut off $A"G" = DE$.   Th 3

4. Through G draw $G"I"$ making $\angle AGI =$ Th 23
   $\angle ABC$.

5. GI, when produced, will intersect $\triangle ABC$ exactly   Post 2, Post 6 (iii)
   one more time, at a point "J".

(Steps 6–9 verify that J is between A and C, as drawn.)

6. J is not on AB.   Post 1

(The reasoning behind step 6 is that if GI produced were to swerve back and intersect AB, then the two points G and J would be joined by two different straight lines, in violation of Postulate 1.)

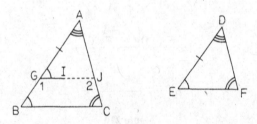

Figure 211

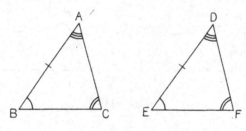

Figure 212

7. GJ is parallel to BC,   4, Th 27/28
8. so J is not on BC.   7, Def "parallel"
9. Therefore J is on AC, between A and C.   6, 8
10. $\angle AGI = \angle DEF$ 4, 1, C.N. 1
11. Triangles AGJ and DEF are congruent.   ASA

(The contradiction appears from an unexpected direction. We will show that
the angle-sum of quadrilateral $BCJG$ is 360°!)

12. $\angle AGI + \angle 1 = 180°$ Th 13
13. $\angle ABC + \angle 1 = 180°$ 4, 12, C.N. 6 (If $a = b$
                                                             and $a + c = d$ then
                                                             $b + c = d$.)

14. $\angle AJI = \angle DFE$ Def "congruent"
15. $\angle AJI = \angle ACB$ 14, 1, C.N. 1
16. $\angle AJI + \angle 2 = 180°$ Th 13
17. $\angle ACB + \angle 2 = 180°$ 15, 16, C.N. 6 (same
                                                             as 13)

18. Therefore the angle-sum of quadrilateral $BCJG$ 13, 17, C.N. 2
    is 360°.
19. But the angle-sum of $BCJG$ is less than 360°.   Corollary to Th H14
20. Contradiction.   18 and 19
21. Therefore $AB = DE$ (see Figure 212).   2–20, logic
22. Therefore triangles ABC and DEF are con-   1, 21, ASA (or AAS)
    gruent.

Our last formal theorems are H16, H17, and H18. Theorem H16 will be
needed to prove Theorem H17, which in turn will set the stage for Theorem
H18. Theorem H18 will explain, finally, why divergent parallels are called
"divergent."

Theorem H16. *If $XABY$ is a biangle and $WCDZ$ is a figure made of three
straight lines such that $CD = AB$, $\angle WCD = \angle XAB$, and $\angle CDZ = \angle ABY$,
then $WCDZ$ is a biangle also.* (See Figure 213.)

Proof.

1. $XABY$ is a biangle and $WCDZ$ is a figure made   hypothesis
   of three straight lines such that $CD = AB$,
   $\angle WCD = \angle XAB$, and $\angle CDZ = \angle ABY$.

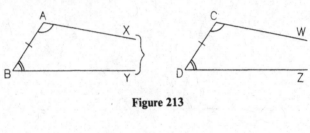

Figure 213

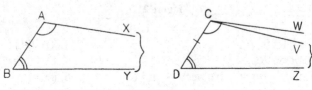

Figure 214

(To show: $WCDZ$ is a biangle also, i.e., that CW and DZ are a-parallel in the direction away from CD.)

2. Pretend CW and DZ are not a-parallel in the direction away from CD.	RAA hypothesis
3. Through C draw C"V" so that CV and DZ are a-parallel in the direction away from CD. Say, to be specific, that CV enters $\angle WCD$.	Post H

(The argument would be the same if CW entered $\angle VCD$. See Figure 214.)

4. $VCDZ$ is a biangle.	Def "biangle"
5. $\angle VCD = \angle XAB$	AB
6. Therefore $\angle WCD = \angle VCD$.	1, 5, C.N. 1
7. But $\angle WCD > \angle VCD$.	C.N. 5
8. Contradiction.	6 and 7
9. Therefore CW and DZ are a-parallel in the direction away from CD.	2–8, logic
10. Therefore $WCDZ$ is a biangle.	Def "biangle"

If two straight lines have a common perpendicular, as in Figure 215, then they are d-parallel by Theorem H9. (This is how we proved Theorem H12.) The next theorem gives us the converse: If two straight lines are d-parallel, then they have a common perpendicular. The theorem goes on to say that the

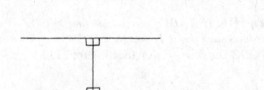

Figure 215

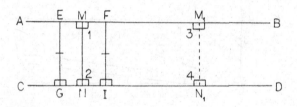

Figure 216

common perpendicular is "unique," i.e., that there is only one. The proof is by Hilbert, and it is quite involved.

Theorem H17. *Two divergent parallels have a unique common perpendicular.*

Proof.

1. Let "*AB*" and "*CD*" be *d*-parallels (see Figure 216). — hypothesis

2. Choose points "*E*" and "*F*" at random on *AB*, and draw *E*"*G*" and *F*"*I*" perpendicular to *CD* (or *CD* produced). — Post 10, Th 12 (Post 2)

Case 1. EG = FI.

3. Then *GIFE* is a Saccheri quadrilateral with summit *EF*. — Def "Saccheri quad."

4. Bisect *EF* at "*M*", *GI* at "*N*", and draw *MN*. — Th 10, Post 1

5. *MN* is a common perpendicular to *AB* and *CD*. — Corollary to Th A

6. Pretend *AB* and *CD* have another common perpendicular "$M_1 N_1$". — RAA hypothesis

7. Since $\angle 1 = 90°$, $\angle 2 = 90°$, $\angle 3 = 90°$, and $\angle 4 = 90°$, — 5, 6

8. the angle-sum of quadrilateral $NN_1 M_1 M$ is 360°. — 7, C.N. 2

9. But the angle-sum of $NN_1 M_1 M$ is less than 360°. — Corollary to Th H14

10. Contradiction. — 8 and 9

11. Therefore *MN* is the only common perpendicular to *AB* and *CD*. — 6–10, logic

(Of course, it is very unlikely that *EG* = *FI*—*E* and *F* were chosen at random. The real work is in Case 2.)

Case 2. EG ≠ FI, say EG > FI. (See Figure 217.)
(If *FI* were the longer the argument would be similar. Case 1 has shown us how easy it is to get a common perpendicular, and to prove it is unique, once we have two perpendiculars of equal length. Our strategy, then, will be to locate two such perpendiculars.)

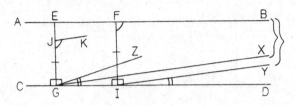

Figure 217

12. From *EG* cut off "*J*"*G* = *FI*.   Th 3
13. Draw *J*"*K*" making ∠*KJG* = ∠*BFI*.   Th 23
(To carry out our strategy, we need to show *JK* produced will intersect *AB*, or *AB* produced. This will take a long time—17 steps!)
14. Draw *G*"*X*" and *I*"*Y*" *a*-parallel to *AB* in the   Post H
    direction toward *B*.
(*GX* is above *CD*, as drawn, because *GX* is *a*-parallel to *AB* through *G* and *CD* is *d*-parallel to *AB* through *G*. Similarly *IY* has been correctly drawn.)
15. Draw *G*"*Z*" making ∠*ZGD* = ∠*YID*.   Th 23
(Steps 16–20 show that *GZ* is above *GX*, as drawn.)
16. *GX* and *IY* are *a*-parallel to one another in the   Th H5
    same direction in which they are *a*-parallel to
    AB.
17. Therefore *XGIY* is a biangle.   Def "biangle"
18. ∠*YID* > ∠*XGD*   Th H8
19. ∠*ZGD* > ∠*XGD*   15, 18, C.N. 6 (If
                                                    a = *b* and *b* > *c* then
                                                    a > *c*.)
20. *GZ* enters ∠*EGX*, as drawn.   19
21. Therefore *GZ* is not parallel to *AB*.   14, Post H (3)
22. Produce *GZ* until it intersects *AB*, or *AB* pro-   Post 2
    duced, at a point "*L*".
(Our immediate goal, remember, is to show *JK* produced intersects *AB*. We have just shown it is surrounded by a triangle, △*EGL*. *JK* produced can't intersect side *EG* a second time, so if only we can show it won't intersect side *GL*, we will be able to conclude it intersects side *EL* and so *AB*. To show *JK* produced won't meet side *GL*, we will show *KJGL* is a biangle. Here's where Theorem H16 comes in.)
23. ∠*JGD* = ∠*FID*   2, Post 4
24. ∠*JGL* = ∠*FIY*   23, 15, C.N. 3
25. *KJGL* is a figure made of three straight lines   13, 12, 24
    such that ∠*KJG*, *JG*, and ∠*JGL* are equal,
    respectively, to ∠*BFI*, *FI*, and ∠*FIY* of biangle
    BFIY.
26. Therefore *KJGL* is a biangle also.   Th H16

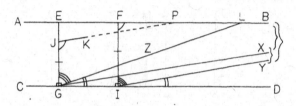

Figure 218

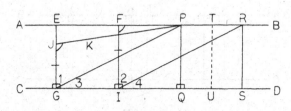

Figure 219

27. Therefore *JK* and *GL* are *a*-parallel.   Def "biangle"
28. *JK* produced intersects △*EGL* exactly one more   Post 2, Post 6 (iii)
    time, at a point "*P*".
29. *P* is not on *GL* or *EG*.   27, Post 1
30. Therefore *P* is on *EL*, as drawn (see Figure 218).   28, 29
(Now that we finally have *JK* produced intersecting *AB*, we will set about
constructing our equal perpendiculars.)
31. Draw *P"Q"* perpendicular to *CD*, or *CD* pro-   Th 12 (Post 2)
    duced.
32. From *FB*, or *FB* produced, cut off *F"R"* = *JP*;   (Post 2) Th 3, Post 1
    from *ID*, or *ID* produced, cut off *I"S"* = *GQ*;
    draw *RS*, *PG*, and *RI* (see Figure 219).
33. Triangles *JGP* and *FIR* are congruent.   SAS
34. *PG* = *RI* and ∠1 = ∠2   Def "congruent"
35. ∠3 = ∠4   23, 34, C.N. 3
36. Triangles *PGQ* and *RIS* are congruent.   SAS
37. *RS* is perpendicular to *CD*, and *PQ* = *RS*.   36, Def "congruent"
(Now we're essentially done, because we can reason with *PQ* and *RS* as we did
with *EG* and *FI* in Case 1.)
38. Bisect *PR* at "*T*", *QS* at "*U*", and draw *TU*.   Th 10, Post 1
39. *TU* is perpendicular to both *AB* and *CD*, and is   imitate proof of Case 1
    the only common perpendicular *AB* and *CD*
    have.

Theorem H18. *Divergent parallels veer away from each other on either side of
their common perpendicular.*

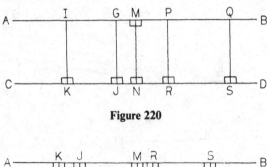

Figure 220

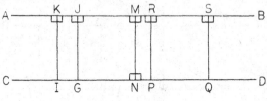

Figure 221

(Divergent parallels, in a word, diverge. What the theorem is saying, specifically, is that the common perpendicular represents the closest approach of the *d*-parallels to one another; and that the further from the common perpendicular a point on either *d*-parallel is chosen, the longer the perpendicular to the other *d*-parallel will be. In Figure 220, *G* is any point on *AB* to the left of the common perpendicular *MN*, and *I* is any point to the left of *G*; *P* is any point to right of *MN*, and *Q* any point to the right of *P*; from these points perpendiculars have been drawn down to *CD*. In Figure 221, *G*, *I*, *P*, and *Q* have been taken on *CD* instead, and perpendiculars drawn up to *AB*. What we have to prove is that, in either figure, $IK > GJ > MN$ and $MN < PR < QS$. By symmetry it will be enough to prove $MN < PR < QS$ in Figure 220—the other three arguments are similar.)

Proof. (Exercise! I thought you might find it fun to prove our last theorem.)

Glimpses

We could go on, indefinitely. There are hundreds of theorems in hyperbolic geometry, with new ones being added all the time. But we have to break off somewhere.

Before we leave the subject altogether, though, I'd like to point out some interesting features of the hyperbolic landscape that can be glimpsed from where we stand now.

First: it's time to reconsider the way we draw straight lines.

For a long time our drawings looked "wrong" in only one respect—they suggested that parallel lines would meet, if only they were produced a bit

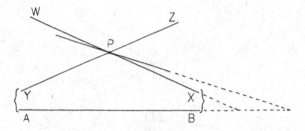

Figure 222

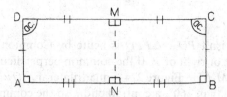

Figure 223

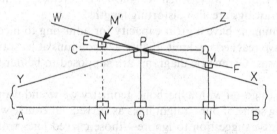

Figure 224

further (Figure 222). But we learned to ignore this suggestion, buoyed by the fact that the drawings looked "right" in every other respect—through Theorem H12, in fact, every new thing we learned was at least consistent with, and often displayed explicitly in, our drawings.

Beginning with Theorem H13, however—"the summit angles of every Saccheri quadrilateral are acute"—many of our drawings have also looked "wrong" in another respect—angles. Figure 223 is our standard representation of a Saccheri quadrilateral, with everything we know indicated. The right angles do look right, but so do the acute angles! Of course we could choose to ignore this suggestion, too, or to think of angles D and C as smaller than right angles by an imperceptible amount. But what Theorems H17 and H18 have told us is much harder to shoo away.

We knew for a long time that *some* pairs of *d*-parallels have a common perpendicular. CD and AB do, in Figure 224. What's new—this is Theorem H17—is that *every* pair does. In particular, then, EF and AB have a common

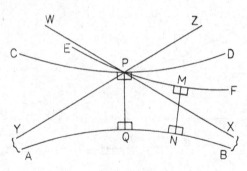

Figure 225

perpendicular. It isn't PQ—$\angle FPQ$ is acute by Common Notion 5—so it must be to the left or right of P. If the common perpendicular to EF and AB were to the left—$M'N'$ in Figure 224—quadrilateral $N'QPM'$ would have an angle-sum greater than 360°, a contradiction; so the common perpendicular must be, in fact, to the right of P, where I've sketched it (MN). But now the angles at M don't *look* right. Furthermore, it looks as if EF keeps getting closer to AB to the right of MN, when we know (Theorem H18) it gets further away—our drawings are also distorting lengths!

While we humans have a great capacity for adjusting to inconvenience, I personally have reached my limit. Figure 224 contains at least three separate false suggestions. Geometric diagrams are supposed to be our servants, not our masters.

If we were to go on with hyperbolic geometry we would want, I think, to begin drawing some of our straight lines as curved. Of course we would then have a new false suggestion to ignore—those curved lines would represent *straight* lines—but that would be the *only* suggestion we would have to ignore. If Figure 224 were redone in this new style the result would be Figure 225, in which the properties of parallels, angles, and lengths are all represented accurately.

Second: the Theorem of Pythagoras, the climax of our study of Euclidean geometry, is false in hyperbolic geometry.

This is no great surprise, as we proved the Theorem of Pythagoras (Theorem 47) using Postulate 5. But with Theorem H13 we have gone far enough in hyperbolic geometry to see *why* the Theorem of Pythagoras is false.

Figure 226 is a "new-style" rendition of Figure 223—a Saccheri quadrilateral with everything we know, including Theorem H13, indicated. It looks as if DC is longer than AB, doesn't it? It *is* longer! And I will leave it as an exercise for you to prove that, indeed, the summit of a Saccheri quadrilateral is always longer than the base.

In Figure 227, $\triangle ABC$ is a right-angled triangle, on which a Saccheri quadrilateral has been constructed by Theorem B. If the Theorem of Pyth-

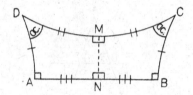

Figure 226

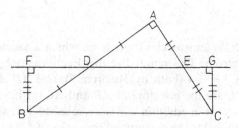

Figure 227

agoras were true (RAA hypothesis), we could apply it to each of the right-angled triangles $\triangle ABC$ and $\triangle ADE$, yielding

$$BC^2 = AB^2 + AC^2, \quad \text{and} \tag{1}$$

$$DE^2 = AD^2 + AE^2. \tag{2}$$

Now $AD = \frac{1}{2}AB$ and $AE = \frac{1}{2}AC$, so equation (2) can be rewritten

$$DE^2 = (\tfrac{1}{2}AB)^2 + (\tfrac{1}{2}AC)^2$$

$$DE^2 = \tfrac{1}{4}AB^2 + \tfrac{1}{4}AC^2$$

$$DE^2 = \tfrac{1}{4}(AB^2 + AC^2).$$

But $\frac{1}{4}(AB^2 + AC^2) = \frac{1}{4}BC^2$ by equation (1), so

$$DE^2 = \tfrac{1}{4}BC^2 \quad \text{or}$$

$$DE = \tfrac{1}{2}BC. \tag{3}$$

The rub is that DE is also equal to $\frac{1}{2}FG$ (Theorem B), implying that the summit BC is equal to the base FG, a contradiction.

Third: perfect scale models are impossible in hyperbolic geometry.

What we mean by a "scale" model of some object—a house, say—is a model with "the same shape" but different size; and when we say the model has "the same shape" as the original, we mean that corresponding angles are equal and corresponding sides are in proportion.

All our work has been in plane geometry, so let's stick to two dimensions. In

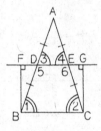

Figure 228

Figure 228 △*ABC* is an isosceles triangle on which a Saccheri quadrilateral has been constructed by Theorem B. In either Euclidean or hyperbolic geometry ∠1 = ∠2, ∠3 = ∠4 (both by Theorem 5), and *AD/AB = AE/AC* = ½ (because *D* and *E* are the bisectors of *AB* and *AC*). I have indicated these things in the figure. If in addition ∠3 were equal to ∠1, all three angles of △*ADE* would be equal to the corresponding angles of △*ABC*; and if the third ratio, *DE/BC*, were also equal to ½, all the corresponding sides of △*ADE* and △*ABC* would be in proportion. Whether △*ADE* is, or is not, a scale model of △*ABC* depends, therefore, on the truth or falsity of two equations:

$$\angle 3 = \angle 1, \quad \text{and}$$

$$\frac{DE}{BC} = \frac{1}{2}.$$

In Euclidean geometry it's easy to prove both equations are true, so △*ADE* *is* a scale model of △*ABC*.

But in hyperbolic geometry neither equation is true, so △*ADE* is *not* a scale model. Right off the bat we can say ∠3 ≠ ∠1: if ∠3 were equal to ∠1 the triangles would be congruent by AAA (Theorem H15), which they are not. In fact ∠3 is *larger* than ∠1: ∠5 = ∠6, making ∠5 + ∠1 half the angle-sum of quadrilateral BCED, so ∠5 + ∠1 < 180° by the Corollary to Theorem H14; but ∠5 + ∠3 = 180°, so ∠3 > ∠1.

In hyperbolic geometry the other equation (*DE/BC* = ½) is false, too. It is equivalent to equation (3) in our discussion of the Theorem of Pythagoras, and leads to the same contradiction.

Here's another example (Figures 229(a) and 229(b)). In either geometry a head-on photograph of a big equilateral triangle (△*ABC*) is a smaller equilateral triangle (△*DEF*). In either geometry, therefore, corresponding sides are in proportion: *DE/AB = DF/AC = EF/BC*.

Now if the universe were Euclidean, angles *D, E, F* would equal angles *A, B, C* and △*DEF* would be a scale model of △*ABC*. But if the universe were hyperbolic we would have Theorem H15 (AAA), by which the corresponding angles *could not* be equal (if they were, triangles *DEF* and *ABC* would be congruent, a contradiction). So in a hyperbolic universe the corners of △*DEF*

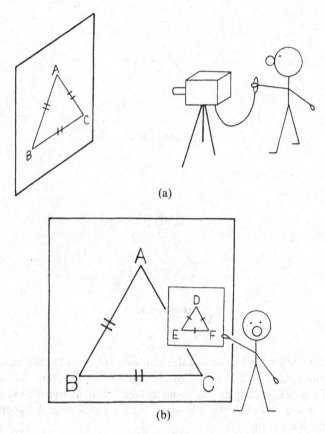

(a)

(b)

Figure 229

would have a different shape than the corners of △*ABC*, and the only way a photographer could get a perfect likeness would be to enlarge his photograph to the size of the original!

Fourth: the larger the area of a hyperbolic triangle, the smaller its angle-sum.

We have glimpsed this already, in our discussion of Figure 228. To describe the phenomenon in general, it will be convenient to introduce the concept of "defect."

Definition H3. The *defect* of a triangle is the amount by which its angle-sum is short of 180°.

Thus, if the triangle is △*ABC* in Figure 230, and we denote its defect by *d*, we have the formula

$$d = 180° - \angle A - \angle B - \angle C.$$

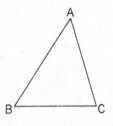

Figure 230

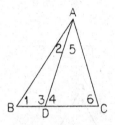

Figure 231

The term "defect" is of course a slur made by fanatical Euclideans. In Euclidean geometry $\angle A + \angle B + \angle C$ would equal 180° (Theorem 32), so d would be 0°—Euclidean triangles "have no defect." But in hyperbolic geometry $\angle A + \angle B + \angle C < 180°$ by Theorem H14, so $d > 0°$ and hyperbolic triangles "have a defect."

To speak of the defect of a triangle is to speak of its angle-sum from another point of view. A triangle with angle-sum 177° has a defect of 3°; one with angle-sum 179.8°, a defect of 0.2°. The advantage of defects is that they are "additive," while angle-sums are not. To see what I mean, choose a point "D" at random on BC, and draw AD (Figure 231). This cuts $\triangle ABC$ into two smaller triangles. If we take the defects of the smaller triangles and add them—

defect of $\triangle ABD$ + defect of $\triangle ADC$

$$= 180° - \angle 1 - \angle 2 - \angle 3 + 180° - \angle 4 - \angle 5 - \angle 6$$

$$= 180° - \angle 1 - \angle 2 - (\angle 3 + \angle 4) + 180° - \angle 5 - \angle 6$$

$$= 180° - \angle 1 - \angle 2 - 180° + 180° - \angle 5 - \angle 6$$

$$= 180° - \angle 1 - (\angle 2 + \angle 5) - \angle 6$$

$$= 180° - \angle 1 - \angle BAC - \angle 6$$

—we get the defect of $\triangle ABC$! It can be proven that this happens no matter how many small triangles we cut a triangle into, and no matter how we make

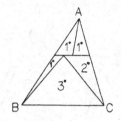

Figure 232

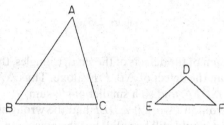

Figure 233

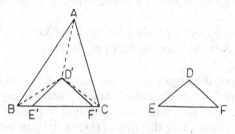

Figure 234

the cuts. In Figure 232, for example, where the numbers within the small triangles are their defects, we can add those defects to conclude that △*ABC* has a defect of 8° (for an angle-sum of 172°). Note that the angle-sums themselves are *not* additive in this way—adding the angle-sums of the small triangles would give 179° + 179° + 179° + 178° + 177° = 892°, a far cry from △*ABC*'s angle-sum of 172°.

The fact that defects are additive is of course the reason triangles with larger areas have smaller angle-sums. We are not in a position to even sketch a complete proof of this, but we can see how the argument would go for one special case. Say △*ABC* and △*DEF* in Figure 233 have different areas, △*ABC* having the larger. If in addition—this is the special case—we could fit a congruent copy of △*DEF* within △*ABC* (△*D'E'F'* in Figure 234), we could by drawing extra straight lines completely partition △*ABC* into small triangles, one of which would be △*D'E'F'*. Since, as we have seen, the defect of △*ABC*

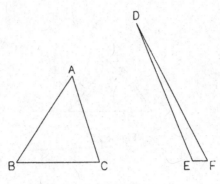

Figure 235

would then be the sum of the defects of the small triangles, the defect of $\triangle ABC$ would be larger than the defect of $\triangle D'E'F'$ alone. Thus $\triangle ABC$ would have a larger defect than $\triangle DEF$, and so a smaller angle-sum.

This argument wouldn't work if $\triangle DEF$ had the wrong shape, as in Figure 235. The conclusion would still be valid, but the reasoning long and difficult.

From the fact that triangles with larger areas have smaller angle-sums, it is possible to deduce an interesting generalization of Theorem H15 (AAA):

If the three angles of one triangle have the same sum as the three angles of another triangle, then the triangles have equal areas.

In the hypothesis the angles no longer need to be equal individually, only have the same sum. The conclusion is weakened accordingly: we can no longer conclude the triangles are congruent, only that they have equal areas. You might like to try establishing this as an exercise. (If you would, remember to take it as known that triangles with larger areas have smaller angle-sums.)

Last: in hyperbolic geometry there is an upper limit for the areas of triangles.[6]

Suppose we were to cut a triangle into a number of equal-defect pieces. For simplicity I've taken $\triangle ABC$, in Figure 236, to be equilateral, so I know its

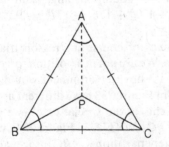

Figure 236

three sides are equal and (by Theorem 5) its three angles are equal. I've bisected $\angle B$ and $\angle C$—these straight lines meet at a point P. I've drawn AP. I know it is easy to prove AP bisects $\angle A$, so I've indicated that also. Since the three small triangles are now congruent (by ASA) they have equal angle-sums and so equal defects; because defects are additive, these defects add up to the defect of $\triangle ABC$; therefore the defect of any one of the small triangles— $\triangle PBC$, say—is exactly 1/3 the defect of $\triangle ABC$.

But the areas have the same relationship: the area of $\triangle PBC$ is exactly 1/3 the area of $\triangle ABC$. This is because the three congruent triangles have equal areas (by Postulate 9).

What I have learned is that

$$\frac{\text{area of } \triangle ABC}{\text{area of } \triangle PBC} = \frac{\text{defect of } \triangle ABC}{\text{defect of } \triangle PBC},$$

which can be rewritten[7]

$$\frac{\text{area of } \triangle ABC}{\text{defect of } \triangle ABC} = \frac{\text{area of } \triangle PBC}{\text{defect of } \triangle PBC}. \tag{1}$$

I can say more. If $\triangle DEF$ (Figure 236) is any other triangle, anywhere in the plane, whose defect is also 1/3 the defect of $\triangle ABC$, $\triangle DEF$ will have the same defect and angle-sum as $\triangle PBC$ and therefore, by the generalization of Theorem H15 on p. 226, the same *area* as $\triangle PBC$. Substituting "defect of $\triangle DEF$" for "defect of $\triangle PBC$" and "area of $\triangle DEF$" for "area of $\triangle PBC$" then gives me

$$\frac{\text{area of } \triangle ABC}{\text{defect of } \triangle ABC} = \frac{\text{area of } \triangle DEF}{\text{defect of } \triangle DEF}. \tag{2}$$

The circumstances under which I've derived equation (2) have been very special: $\triangle ABC$ has been equilateral and—while $\triangle DEF$ itself has had no particular shape—the ratio (defect of $\triangle DEF$/defect of $\triangle ABC$) has had the particular value 1/3. It can be proven, however, that equation (2) remains valid no matter what the shape of $\triangle ABC$, and no matter what the value of the ratio—equation (2) is true for *every* pair of hyperbolic triangles!

Accepting equation (2), we can derive a formula for the area of any triangle. Think of $\triangle ABC$ as a general (variable) triangle whose area we want to find, and $\triangle DEF$ as a particular (fixed) triangle. Let k stand for the ratio (area of $\triangle DEF$/defect of $\triangle DEF$). Substituting k for the right side of equation (2) then gives us

$$\frac{\text{area of } \triangle ABC}{\text{defect of } \triangle ABC} = k,$$

which can be rewritten

$$\text{area of } \triangle ABC = k(\text{defect of } \triangle ABC); \tag{3}$$

this is our formula, valid for every triangle ABC.

k is what mathematicians call a "constant"—it is a number whose value is independent of the particular triangle DEF we originally used to express it. If $\triangle D'E'F'$ had been any other triangle, (area of $\triangle D'E'F'$/defect of $\triangle D'E'F'$) would have been equal to (area of $\triangle DEF$/defect of $\triangle DEF$) by equation (2), and so (area of $\triangle D'E'F'$/defect of $\triangle D'E'F'$) would have given us the same value for k. It's true, of course, that we don't know what the actual value of k is[8]; our situation is analogous to that of Euclid, who was in possession of the formula

circumference of a circle = π(diameter of the circle)

without knowing the exact value of the constant π. But just as Euclid was able, nonetheless, to put his circle-formula to good use, our formula (3) is all we need to prove that areas of hyperbolic triangles have an upper limit.

In hyperbolic geometry, just as in Euclidean geometry, it is impossible for any angle of a triangle to be $0°$. The angle-sum of a triangle, therefore, is always *positive*, so the defect

$$d = 180° - \angle A - \angle B - \angle C$$

of a triangle ABC is always *less* than $180°$. This means that the right side of formula (3) is always less than $k \cdot 180°$, giving us

$$\text{area of } \triangle ABC < k \cdot 180°. \tag{4}$$

As (4) is true for *every* triangle ABC, we see that $k \cdot 180°$ is an upper limit for the areas of hyperbolic triangles. No triangle can have an area exceeding, or even equal to, that fixed number.

Our conclusion will seem less implausible if we draw a series of "new-style" triangles—sides curving inward—as in Figure 237. While their areas do increase, they do not do so at the same rate they would in Euclidean geometry. This enables our imaginations to admit that, just as the increasing series of numbers

$$1/2,\ 3/4,\ 7/8,\ 15/16, \ldots$$

has an unattainable upper limit of 1, it is possible the series of areas has an unattainable upper limit too.

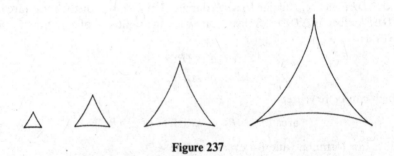

Figure 237

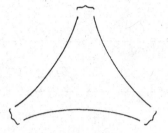

Figure 238

The significance of the particular number $k \cdot 180°$ is that it is the area of the object shown in Figure 238, a nontriangle toward which the triangles in Figure 237 seem to be tending.

Exercises

1. Prove that the angle-sum of every biangle is less than 180°.

2. In Figure 239, $XABY$ is a biangle with $\angle XAB = \angle ABY$. From the midpoint, C, of AB, and in the direction of parallelism, CZ has been drawn at right angles to AB. Prove that CZ is asymptotically parallel to AX and BY.

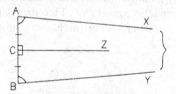

Figure 239

3. In Figure 240, $WABX$ and $YCDZ$ are biangles with $AB = CD$, $\angle WAB = \angle ABX$, and $\angle YCD = \angle CDZ$. Prove that all four angles are equal.

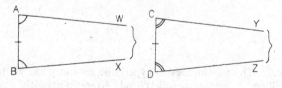

Figure 240

4. It is given, in Figure 241, that EI causes all eight angle relationships (page 84) to occur. If this were Euclidean geometry it would follow that all eight angle relationships occur at JM as well. Prove that in hyperbolic geometry, on the contrary, *none* of the eight angle relationships occurs at JM.

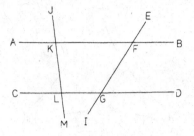

Figure 241

5. In Euclidean geometry the straight line joining the midpoints of two sides of a triangle is equal to half the third side—in Figure 242, $DE = \frac{1}{2}BC$. Prove that in hyperbolic geometry, $DE < \frac{1}{2}BC$.

Figure 242

6. In Euclidean geometry an angle inscribed in a semicircle ($\angle BAC$ in Figure 243; Q is the center of the circle) is equal to 90°. Find the relationship between $\angle BAC$ and 90° in hyperbolic geometry, and prove that your answer is correct.

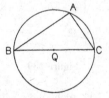

Figure 243

Notes

[1] *Definition H1.* The terminology in hyperbolic geometry has yet to settle down. What we are calling "asymptotic parallels" and "divergent parallels" are given various names in other books.

[2] *a biangle* is known by various names in other books. See the previous note.

[3] *the 89.9999999995° value we mentioned before.* Though our argument suggests this conclusion it does not really prove it, because for all we have said the increasing sequence

$$\angle 1 < \angle 2 < \angle 3 < \angle 4 < \cdots$$

might have values like

$$82° < 82.9° < 82.99° < 82.999° < \cdots$$

that stay below some value—in this case, 83°—that is smaller than 90°. Nonetheless our conclusion *can* be proven, and it *is* correct.

[4] *two million light-years long.* At the other extreme "short" might mean "two zillionths of a millimeter," in which case $\angle XPQ$ would be noticeably less than a right angle in drawings the size of those in this book. Letting "*a*" be any fixed acute angle— and it will suit our purpose to think of *a* as having a value that can be distinguished from a right angle with existing instruments, say $a = 89°$—there are actually an infinite number of conceivable hyperbolic geometries, corresponding to the different values that could be required of PQ to make $\angle XPQ = a$. It is conceivable for example that $\angle XPQ = 89°$ when PQ is only a meter long. Such a hyperbolic geometry could not possibly be true, for surveyors would long ago have noticed so drastic a variance from the geometry of Euclid. But it is equally conceivable (this is my point) that $\angle XPQ$ will not take on the value 89°, or any other value detectably less than 90°, until PQ is so extremely long as to be far beyond the range of human experience; and *that* kind of hyperbolic geometry might well, for all we know, be true.

[5] *little prejudice concerning them.* Similarly one's common sense can accept the possible truth of Einstein's Special Relativity, for its effects (the shrinking of moving rods, the slowing of moving clocks) are detectable only in circumstances outside those of daily life. If you have studied much physics you have probably noticed that the relation hyperbolic geometry would have to Euclidean geometry in my hypothetical universe is reminiscent of the relation Einsteinian mechanics has to Newtonian mechanics.

[6] *upper limit for the areas of triangles.* Remember Gauss' Postulate (p. 128)?

[7] *which can be rewritten.* In equation (1) we seem to be dividing apples by oranges, because areas are measured in square units and defects in degrees. But this can be overcome with a different system of measurement wherein areas and defects are both expressed as pure numbers.

[8] *we don't know what the actual value of k is.* Indeed, we cannot know. There are infinitely many conceivable hyperbolic geometries, each with a different value of k.

CHAPTER 7

Consistency

At the beginning of Chapter 5 we argued for the mathematical legitimacy—the word we used was "consistency"—of hyperbolic geometry. (Recall that an axiomatic system is *consistent* if no contradiction can be deduced from its foundation of primitive terms, defined terms, and axioms.) Our case was based on two assumptions—one explicit, the other implicit.

The explicit assumption was

(1) Neutral geometry by itself does not imply Postulate 5

(which we also numbered "(1)" at the time—page 154). We found this a plausible assumption to make in view of the work we had done in Chapter 4. We said we would verify it later on (i.e., now).

The implicit assumption was

(2) Neutral geometry is consistent.

You may not have noticed this assumption at the time. I didn't play it up, because our discussion was already quite subtle and I didn't want to add a further complication. It had probably never occurred to you that Neutral geometry, or for that matter Euclidean geometry, might be anything but consistent. Certainly from the very beginning of our study of the *Elements*, long before we ever used the word "consistent," we had always taken it for granted that Neutral geometry and Euclidean geometry were consistent. For that matter no mathematician has ever, as far as I know, expressed serious doubt on this point.

Still, (2) *is* an assumption, and one that was crucial to our case. (This is easy to see. The hyperbolic foundation includes the Neutral foundation, so any contradiction deducible from the latter would also be deducible from the former.) Thus we should address ourselves to assumption (2) as well as to (1).

In fact we should address ourselves to the stronger assumption

(2') Euclidean geometry is consistent,

because historically that's the assumption that was made, even though strictly speaking our case only required (2). (2') includes (2).

Our goal at the beginning of Chapter 5 was to establish that

(3) hyperbolic geometry is consistent.

We deduced this from the previous assumptions, because that's how people first came to believe that a non-Euclidean geometry was possible. They started out believing in (1) and (2′)—and therefore in (2)—and had the perceptiveness and courage to follow these assumptions to (3).

But there are practical objections to keeping that organization now. For one thing neither (2′) nor (2) has ever been exhaustively verified, and there is reason to believe that neither one ever will be. For another thing, even if we accept (2′), say, as a background hypothesis, what's going on will be a lot clearer if we verify (3) first and use it to prove (1), rather than the other way around. (1) is still of interest in itself, because it settles the centuries-old question about Postulate 5.

So the plan is this. For now we will continue to assume that (2′) is correct—and, as I said, there is no good reason for suspecting it is not. We will use (2′) to prove (3). Then we will use (3) to prove (1). Later in the chapter we will return to (2′) and say what we can about it.

If this chapter sounds too weighty for your taste, you can skip most of it. The next one will still be comprehensible. But do read "Poincaré's Model" starting on page 235—it gives a delightful picture of what a hyperbolic universe might look like "from the outside." Between that and the earlier section "Reconciliation With Common Sense" (page 207) you'll be able to cultivate an intuitive "feel" for hyperbolic geometry that will be strong enough to dispel any serious doubts you may have about its consistency. The reason we have never doubted the consistency of *Euclidean* geometry is that we have had a clear intuitive picture of it from the very beginning.

Models

If we are going to prove

(3) hyperbolic geometry is consistent,

we should first consider how that might be done.

Remember that, officially, we are still considering hyperbolic geometry to be a formal axiomatic system. (The section "Reconciliation With Common Sense" was an unofficial aside.) Remember too that in a formal axiomatic system the primitive terms have no meanings.

When it is claimed that a formal axiomatic system is consistent, the evidence that is usually offered is what is known as a "model" for it.

Definition. A *model* for a formal axiomatic system is an interpretation of the primitive terms under which the axioms become true statements.

Here the word "interpretation" is not used in its usual sense of "clarification of meaning"—there is no meaning to clarify—but in a more basic sense of "giving of meaning." The idea is that since the primitive terms have *no* inherent meanings, we can assign them any meanings we choose. Doing this makes the axioms meaningful as well; if in fact it makes all of them *true*, we have a model.

Here, for example, is a model for "The Scorpling Flugs," the little formal axiomatic system we considered back in Chapter 5 ("A Simple Example of a Formal Axiomatic System," page 164).

Make a stack of four books on the floor next to your chair. These will be the four "flugs." We will say that one "scorples" another if it is higher than the other in the stack. Displaying this schematically we get a "dictionary"—

primitive term	*interpretation*
the flugs	the four books in the stack
to scorple ...	to be higher than ...

—which we can use to translate the axioms into statements about the stack of books:

SF1. If A and B are distinct books in the stack, then A is higher than B or B is higher than A.

SF2. No book in the stack is higher than itself.

SF3. If A, B, and C are books in the stack such that A is higher than B and B is higher than C, then A is higher than C.

SF4. There are exactly four books in the stack.

Since all these statements are obviously true, the interpretation is a model as claimed.

Whenever we have a model, all the theorems translate into true statements as well. In the present case, for example, Theorem SF3 becomes "There is at least one book in the stack that is higher than every other book in the stack," which is clearly true. (The book of Theorem SF3, which is proven to be unique in Theorem SF4, is dubbed the "pushy" book in Definition SF1. Thus the pushy book is the top book.)

We conclude from our model that The Scorpling Flugs is a consistent system. This is because any contradiction deducible from Axioms SF1–SF4 would translate into a contradiction about the stack of four books, and we take it for granted that there are no contradictions in the physical world.

Here is another model of The Scorpling Flugs, one that provides a less satisfactory proof of consistency. (Having seen the previous model there can be no doubt that The Scorpling Flugs is consistent, but I am speaking as if that model had not been presented.) When you look at the "dictionary" below, don't panic at the combinations of symbols in the second column—all you will need to know about them is that each combination represents a number.

primitive term	*interpretation*
the flugs	the four symbol-combinations $2^{\sqrt{3}}$, tan 1.28, $\lim_{n \to \infty}(1 + (1/n))^n$, and $\int_1^6(1/x)dx$.
to scorple ...	to represent a number smaller than the number represented by ...

Under this interpretation it is less obvious that the statements into which the axioms are transformed are true. For example Axiom SF1 translates into

If A and B are distinct symbol-combinations drawn from the four given symbol-combinations, then either A represents a number smaller than the number represented by B, or B represents a number smaller than the number represented by A.

Is that true? Sometimes in mathematics it *looks* like we have two different numbers, when what we actually have are only two different representations of the same number—like $|\sqrt{81}|$ and 3^2, that both represent 9. For all we can tell at the moment, this may be true of, say, $2^{\sqrt{3}}$ and tan 1.28.

As a matter of fact the translated axioms *are* all true (in particular $2^{\sqrt{3}}$ is slightly less than tan 1.28), but considerable calculation is required before that becomes clear. And since this labor is performed using the axioms and theorems of *another axiomatic system* (called "Real Analysis"), we have only *transferred* the consistency question from The Scorpling Flugs to Real Analysis. (Real Analysis is the giant axiomatic system that includes calculus, trigonometry, coordinate geometry, and high-school algebra.[1]) *If* Real Analysis is consistent, then The Scorpling Flugs is consistent—because any contradiction in The Scorpling Flugs would translate into a contradiction in Real Analysis. But unless we can go on to prove the consistency of Real Analysis, we are stuck with that weak-sounding conclusion: The Scorpling Flugs is consistent if Real Analysis is.

We gather from the above that there are different *kinds* of models, and that they indicate consistency to different *degrees*. The best models are like the first—simple, physical, "down to earth," able to be grasped down to the last detail. In the face of such a model a system's consistency can scarcely be doubted, and so the model is said to provide an "absolute" proof of consistency. A model like the second, which only transfers the consistency question to another system, is said to provide a "relative" proof of consistency.

Poincaré's Model

Unfortunately, as far as the consistency of hyperbolic geometry is concerned, only relative proofs are known.

The first were due to János Bolyai and Lobachevsky. Like the second model of the previous section, their models reduced the consistency question to that of Real Analysis.

Then, in the last third of the nineteenth century, beginning with a model by Eugenio Beltrami in 1868 (see page 171), a whole series of *geometric* models were constructed.[2] (The news about non-Euclidean geometry had only begun to be widely circulated in the 1860s.) One of these, proposed by the French mathematician, physicist, and philosopher Henri Poincaré (1854–1912), is especially easy to comprehend, and is the one we will examine.

Poincaré's model reduces the consistency of hyperbolic geometry to that of Euclidean geometry. Since we've always taken it for granted that Euclidean geometry is consistent—this is assumption (2′) on page 232—Poincaré's model is pretty convincing. Poincaré's model has the added advantage that, as Poincaré himself did in his book *Science and Hypothesis*[3] (1902), it can be presented in story form, making it possible to gloss over technical points we would otherwise have to spend a long time verifying.

Here we go.

Let "$\mathscr{C}$" be one of Euclid's circles, located somewhere in Euclid's plane. We will suppose that its radius R is large enough for $\mathscr{C}$ to accommodate a sizable population of two-dimensional people living inside. Our perspective will be that of giant, godlike spectators standing on the plane outside of $\mathscr{C}$, watching what happens inside.

$\mathscr{C}$ is filled with a funny gas that causes meter-sticks (sticks one meter long when placed at the center of $\mathscr{C}$) to shrink as they move away from the center. The precise formula that describes this is

$$(\text{length of a meter-stick at distance } r) = 1 - \frac{r^2}{R^2} \text{ meters,}$$

where we use the half-meter mark on the stick to measure the stick's distance r from the center of $\mathscr{C}$. Thus if we were to place the stick successively at the center (so $r = 0$), halfway to the rim ($r = \frac{1}{2}R$), three-quarters of the way to the rim ($r = \frac{3}{4}R$), and so on, and calculate its various lengths from the formula, we would obtain the following table. (Example: when $r = (7/8) R$ the formula yields a length of $1 - [(49/64) R^2]/R^2 = 1 - (49/64) = 15/64$, which is 0.2348 to four decimal places.)

distance from center	length of stick (meters)
0	1.0000
$\frac{1}{2}R$	0.7500
$\frac{3}{4}R$	0.4375
$\frac{7}{8}R$	0.2348
$\frac{15}{16}R$	0.1211
$\frac{31}{32}R$	0.0615

Supposing further that *everything* inside $\mathscr{C}$ (including the people who live there) experiences a corresponding variation of linear dimensions, it follows that no insider is aware of these strange goings-on! A man who at the center of $\mathscr{C}$ is one meter tall, as measured by a meter-stick he carries with him, will, after walking three-quarters of the way to the rim, *still* be one meter tall according to the very same stick (see Figure 244). His surroundings will have

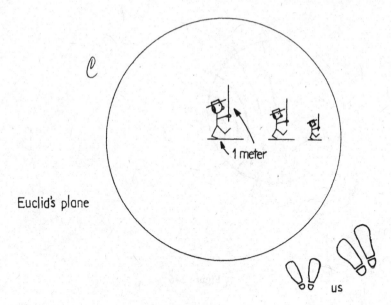

Figure 244

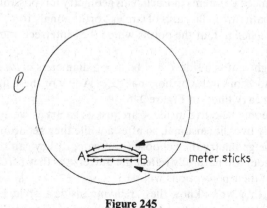

Figure 245

stayed in proportion, so only we outsiders will be aware that the stick, his body, his hat, his stride, as well as trees, cars, etc. are only 0.4375 as long as they used to be.

Finally we will suppose that the gas filling $\mathscr{C}$ causes a ray of light travelling between two interior points to always take the "shortest" path, *as measured by the insiders*. From *our* perspective a light-path joining two points will be a straight line if the two points are on a diameter of $\mathscr{C}$; otherwise the path will bulge toward the center, because meter-sticks get longer as they move in that direction. In Figure 245, for example, the straight-line path from A to B is six meters long, but the curved path is only five meters long.

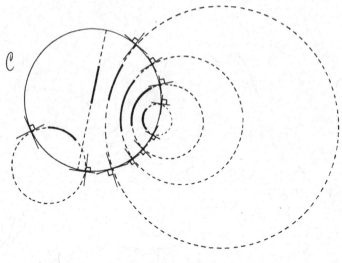

Figure 246

Using advanced theorems in Euclidean geometry it is possible to prove that these curved paths are in fact arcs of circles "orthogonal" to $\mathscr{C}$. Two circles are said to be *orthogonal* if, at the points where they intersect, their tangents are perpendicular.

Thus the light-paths inside $\mathscr{C}$ will be along diameters of $\mathscr{C}$, and also along arcs interior to $\mathscr{C}$ of circles orthogonal to $\mathscr{C}$. A few of these light-paths have been drawn (heavy lines) in Figure 246.

Now the people who live inside $\mathscr{C}$ are just as smart as we are, even if their brains *are* only two dimensional, so after a while they get around to studying geometry. The geometry they choose (or create, if they can't find a suitable one ready-made) will naturally reflect the universe as they perceive it, so let us consider what their perceptions will be.

First of all, they won't know they are living inside a circle. Laying a meter-stick end-over-end on (what *we* know to be) one of $\mathscr{C}$'s radii, no matter how many times, they will never reach the rim because the stick is shrinking too fast. (If you don't see this, think of it another way. Near the rim the length of a meter-stick, along with every other length, is shrinking toward $1 - (R^2/R^2) = 0$, so if an expedition of insiders *were*, by some miracle, to reach the rim, they would never live to tell about it!) Thus to the insiders the interior of $\mathscr{C}$ stretches infinitely far in all directions, and constitutes their "plane."

Second, they will naturally understand a "straight line" to be either the path taken by a ray of light, as Euclid may have done, or as the shortest path between two points, as Archimedes did (page 31). Inside $\mathscr{C}$ the two are equivalent, so in either case the insiders' "straight lines" will be what are, for us, portions interior to $\mathscr{C}$ of diameters of $\mathscr{C}$ and circles orthogonal to $\mathscr{C}$. (Since light-rays follow precisely these paths any of the heavy lines in Figure 246 will

Figure 247. M. C. Escher, *Circle Limit I* (1958). © M. C. Escher Heirs c/o Cordon Art-Baarn-Holland. The Poincaré model packed with flying fish. Their backbones are "straight lines." Escher was inspired to illustrate a hyperbolic universe by conversations with the geometer H. S. M. Coxeter of the University of Toronto. This was Escher's first attempt; see his much more graceful *Circle Limit III* (1959) or *Circle Limit IV* (1960) in, e.g., *The World of M. C. Escher* (Harry N. Abrams, 1971).

look straight to an insider who sights along it.) Full "infinite straight lines" will be diameters minus their endpoints and orthogonal arcs minus their points of intersection with $\mathscr{C}$.

Third, they will accept what we have called "Postulate H." In Figure 248, for example, where A and B lie on a diameter, the "asymptotic parallels" to AB through P will be the two orthogonal arcs YPZ amd WPX passing through the endpoints Y^* and X^* of the diameter (these points are on $\mathscr{C}$ and so do not exist for the insiders). The "divergent parallels" will be the orthogonal arcs joining P to the various points on $\mathscr{C}$ between Y^* and W^*, and the

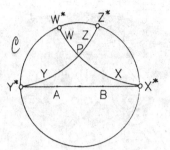

Figure 248

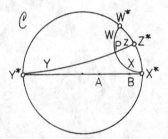

Figure 249

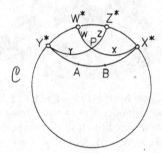

Figure 250

nonparallels will be the orthogonal arcs (plus one diameter) joining P to the points on $\mathscr{C}$ between W^* and Z^*. When P, A, and B are located less symmetrically with respect to the center of $\mathscr{C}$, or when A and B do not lie on a diameter, the situations are entirely analogous. See Figures 249, 250, and 251.

Fourth (and last), the insiders would accept the axioms of Neutral geometry: Common Notions 1–6, Postulates 1–4, and Postulates 6–10. As the Common Notions merely express general principles for reasoning with quantities, it is only to be expected that the insiders would accept these; accordingly, the substance of my statement is that the insiders would accept the

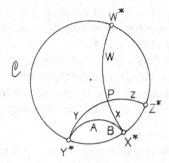

Figure 251

postulates of Neutral geometry. We cannot even begin to verify this without going a *great* deal further in Euclidean geometry than we have, but we can at least see what *kind* of Euclidean theorems a verification would involve.

Consider Postulate 1, which says, "It is possible to draw one and only one straight line from any point to any point." In view of the insiders' notions of "point" and "straight line," this is really (that is, from our perspective) a statement about interior points, diameters, and orthogonal arcs of $\mathscr{C}$. Translating their understanding of Postulate 1 into our terms, we get

(∗) Given two points interior to a fixed circle $\mathscr{C}$, it is possible to draw one and only one diameter of $\mathscr{C}$, or one and only one circle orthogonal to $\mathscr{C}$ (not both), through the two points.

Now it so happens that this is a theorem in advanced Euclidean geometry. Therefore, standing as we are in Euclid's plane, we will find that (∗) is true. Since Postulate 1 is only the insiders' way of expressing that same truth, they will find Postulate 1 to be true.

The same sort of thing happens for all the other postulates of Neutral geometry. Translating them out of the peculiar argot of the people living inside $\mathscr{C}$, every one of them becomes a provable theorem in Euclidean geometry and is therefore true in the larger universe of which the world inside $\mathscr{C}$ is a part. Giving the insiders credit for being as perceptive as we are, they will notice these truths, and so accept all the postulates of Neutral geometry.

(If the last two paragraphs are unclear maybe it would help if I repeated the explanation, this time starting from our perspective.

(The context here is Euclid's plane, in which, because it *is* Euclid's plane, all Euclidean theorems are true. These theorems are true for the insiders just as they are true for us, except that, owing to the effects of the gas inside $\mathscr{C}$, the insiders *express* these truths in different terms. What we call a "point inside $\mathscr{C}$" they call simply a "point"; what we call "the portion interior to $\mathscr{C}$ of a circle orthogonal to $\mathscr{C}$" they call "an infinite straight line"; and so on.

(The heart of the matter is that there are nine provable theorems in advanced Euclidean geometry which the insiders would express as the nine postulates of Neutral geometry, and consequently the insiders will find the postulates of Neutral geometry to be true. Statement (∗) is the theorem the insiders would express as Postulate 1.)

Granting that the insiders will indeed accept the axioms of Neutral geometry, we see that, since they also accept Postulate H, they will select hyperbolic geometry as the geometry describing their world.

Poincaré's model consists,[4] then, of the following "dictionary":

primitive term	*interpretation*
point	point inside a fixed circle $\mathscr{C}$ in Euclid's plane
line	portion interior to $\mathscr{C}$ of a Euclidean line
straight line	portion interior to $\mathscr{C}$ of a diameter of $\mathscr{C}$ or a circle orthogonal to $\mathscr{C}$
plane	interior of $\mathscr{C}$

Under this interpretation the postulates of hyperbolic geometry become theorems of Euclid's geometry. Therefore, using this interpretation, any contradiction deducible from the hyperbolic postulates could be translated into a contradiction deducible from the corresponding Euclidean theorems. As we are assuming that Euclidean geometry is free of contradictions, it follows that hyperbolic geometry is too. Or in other words, hyperbolic geometry is consistent if Euclidean geometry is.

As a consequence we have also settled, finally, the venerable question about Postulate 5. Is Postulate 5 deducible from Neutral geometry? Not if Euclidean geometry is consistent.

The argument goes like this. If Postulate 5 *were* deducible from Neutral geometry, then Postulate 5 would be a theorem in hyperbolic geometry. This theorem would contradict Postulate H, so hyperbolic geometry would be inconsistent. But we have just seen that if Euclidean geometry is consistent, hyperbolic geometry is too. Therefore, if Euclidean geometry is consistent, Postulate 5 cannot be deduced from Neutral geometry.

We have proven what we said we would (page 233), but while on the subject of Poincaré's model let me add one further wrinkle.

In Figure 252, Q is the center of $\mathscr{C}$, A and B are on a diameter, and $Q*Q$ is half the perpendicular diameter. P_1, P_2, P_3, etc. are points on $Q*Q$, each half as far from Q as the previous one (from the perspective of *out*siders). $P_1 X_1$, $P_2 X_2$, $P_3 X_3$, etc. are the corresponding right asymptotic parallels to AB (to outsiders, they are the orthogonal arcs joining those points to X^*). Notice how the "angles of parallelism" (see page 206) $X_1 P_1 Q$, $X_2 P_2 Q$, $X_3 P_3 Q$, etc. are becoming more and more nearly right.

Pursuing this, let us take, in Figure 253, a point P on $Q*Q$ that is extremely close to Q, so close that we require a powerful microscope to distinguish it from Q. The dotted circle on the right is how the region around Q appears through our microscope. PX is the right asymptotic parallel to AB (that is, the orthogonal arc through P and X^*). Angle XPQ is exceedingly close to a right angle.

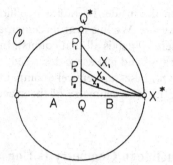

Figure 252

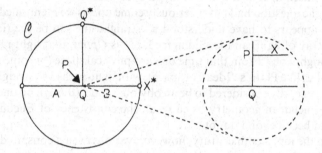

Figure 253

We will suppose, in fact, that we have taken P so near to Q that the difference between angle XPQ and a right angle is beyond the power of the insiders' instruments to detect. We will suppose further that the insiders are very tiny (too small to be seen even through our microscope), that they all live in the vicinity of Q, and, finally, that they have never traveled to, or even seen via telescope, any point further from Q than P.

We have observed that inside $\mathscr{C}$ the "true" geometry—the one actually governing the behavior of the insiders' "points" and "straight lines"—is hyperbolic. When we were speaking earlier we took it for granted that the insiders were large enough, relative to $\mathscr{C}$, to have discovered this for themselves.

With the extra limitations we have just imposed on them, however, their situation is now a two-dimensional analog of the one we imagined *ourselves* to be in, back in "Reconciliation With Common Sense" (page 207). The insiders will find the space around them to be Euclidean! It is really hyperbolic, but they won't be able to detect that, because it will seem to them that angles of parallelism are always right.

This leads to the interesting thought that *we may be three-dimensional versions of the insiders.* Imagine that in Euclidean space there is an enormous sphere painted black on the inside. Its radius is billions of light-years, and contains the known universe. We live near its center, and are of course very tiny (relative to the sphere, which is all that counts). Inside the sphere meter-sticks shrink and light-rays bend just as we described for $\mathscr{C}$. A little reflection will show that these hypotheses are entirely compatible with human experience, yet under them the true geometry of the space around us would be hyperbolic!

Can We Be Sure Euclidean Geometry Is Consistent?

This is what mathematicians asked themselves once the issue of consistency had been raised in connection with hyperbolic geometry.

Before the question had never seriously come up. As we mentioned on page 31, Plato appears to have understood a "straight line" to be the (idealized) path of a ray of light, and Euclid agreed, in his *Optics*, that light rays travel along straight lines. From that time onward physical space (or some idealized version of it, like Plato's "Idea" of space), with light-paths as "straight lines," had been generally considered to be an obvious model (though no one called it that) for Euclidean geometry. As a result the consistency of Euclidean geometry had been beyond doubt.

During the nineteenth century, however, various events conspired to make the matter seem less certain. It became widely recognized that no conceivable experiment could conclusively prove that a Euclidean postulate is true (see "An Experimental Test of Postulate 5," p. 147). Theoretical universes were invented, compatible with human experience, in which Euclidean laws failed to hold (see the previous section). Logical weaknesses were found in the *Elements* itself; though these were successfully shored up with extra axioms and new demonstrations (we did some of this work in Chapter 2), the book's reputation for logical rigor was permanently tarnished.

The possibility that Euclidean geometry might be inconsistent was unthinkable no longer.

An alternative to the traditional model had to be found. The natural choice was "analytic geometry," a part of Real Analysis (*not* of geometry).

It may seem that Real Analysis is too complicated a branch of mathematics within which to build a model of Euclidean geometry. After all, models are supposed to be simple—the consistency of the system that is modeled is reduced to the consistency of the model. But during the nineteenth century the logical structure of Real Analysis had been gone over with a fine-toothed comb, and the whole edifice reerected on a foundation that was particularly simple and solid: the arithmetic of the whole numbers.[5] Consequently by the end of the century mathematicians' confidence in the consistency of Real Analysis was at an all-time high, and that subject was seen to be much more securely grounded than Euclidean geometry.

You encountered analytic geometry—which you may have called "coordinate geometry" or "Cartesian geometry"—in high school, when you drew graphs of equations. You may remember that the graph of an equation like $3x + 4y = 6$ turned out to be a "straight line," or that the graph of $x^2 + y^2 = 4$ was a "circle." At the time these figures were identified, probably without comment, with the "straight lines" and "circles" of Euclidean geometry.

The graph of $3x + 4y = 6$ is not really a Euclidean straight line. At root it is a set of *pairs of real numbers*, specifically the set of all pairs (x, y) of real numbers for which $3x + 4y$ is equal to 6. Calling this set S we have, in symbols,

$$S = \{(x, y): x, y \text{ are real numbers and } 3x + 4y = 6\},$$

which doesn't look *anything like* a straight line.

You may have been lucky enough to have a teacher who pointed this out, and who explained *why*, nevertheless, it is possible to identify S with a Euclidean straight line (and so to illustrate it with the same kind of drawing). The reason is, of course, that there is a standpoint from which a set like S can be seen to have all the *properties* of a Euclidean straight line. And that standpoint constitutes the model mathematicians turned to.

The following "dictionary" makes the interpretation explicit.

primitive term	*interpretation* (all lower-case letters denote real numbers)
point	pair (x, y)
circle[6]	set of all pairs (x, y) that satisfy an equation of the form $(x - h)^2 + (y - k)^2 = r^2$ where h, k, r are fixed and $r > 0$
straight line	set of all pairs (x, y) that satisfy an equation of the form $ax + by = c$ where a, b, c are fixed and a, b are not both 0
the plane	the set of all pairs (x, y)

This interpretation is a model for Euclidean geometry because under it Euclid's postulates are transformed into theorems of Real Analysis. Postulate 1, for example, becomes the following theorem:

Given two pairs (x_1, y_1) and (x_2, y_2) of real numbers, there is one and only one set of the form

$\{(x, y): x, y$ are real numbers and $ax + by = c$ where a, b, c are fixed
     real numbers and a, b are not both 0$\}$

that contains both (x_1, y_1) and (x_2, y_2).

Since the model turns any contradiction in Euclidean geometry into a contradiction in Real Analysis, Euclidean geometry is consistent if Real Analysis is.

There the matter rests. As I write the consistency of Real Analysis is still an open question, and is likely to remain so. Certainly a relative proof of its consistency would be worthless, as Real Analysis has been more thoroughly investigated, and consequently mathematicians trust it more, than any other mathematical system within which a model of it might be built. And because Real Analysis has infinitely many components (the real numbers) and reality, as far as we know, does not (even the number of elementary particles in the known universe is finite), an absolute proof of its consistency by a physical model seems out of the question.

Could there be some other way of showing Real Analysis is consistent, not using a model at all? There are precedents for this. There is, for example, an axiomatic system called "the Propositional Calculus," part of mathematical logic (itself a branch of *mathematics*), for which there is a simple and convincing proof of consistency that does not use a model.[7] (Such a proof is still called "absolute" because it does not assume the consistency of another system.)

Unfortunately it seems that for Real Analysis this avenue too is closed. In 1931 the mathematical logician Kurt Gödel (1906–1978) published a paper in which he showed that, roughly,

it is impossible to establish the logical consistency of any complex deductive system except by assuming principles of reasoning whose own internal consistency is as open to question as that of the system itself.[8]

Here the word "complex" is used to mean that the system is sufficiently complicated to contain the arithmetic of the whole numbers, which Real Analysis certainly is.

Coming on top of a whole series of logical paradoxes that had come to light around the first decade of this century,[9] and which had never been resolved in a completely satisfactory way, the effect of Gödel's paper was to rudely shake mathematicians' confidence in the consistency of mathematical systems generally, and in that of Real Analysis in particular. As I write it remains true that no contradictions have been discovered in Real Analysis, or in any other major branch of mathematics, and that mathematicians continue to behave (they have no choice!) as if none ever will be. But they have lost the innocent hope, which had crested around the turn of the century, that mainstream mathematics will someday be proven secure. Today mathematicians carry on their work in the constant shadow of knowledge that—as one logician put it at mid-century—

[T]here is the awkward possibility that present-day mathematics is actually in serious error, so that any formal system which gives a reasonable facsimile of present-day mathematics must contain a contradiction. We do not believe this to be the case, but we can offer no reason why it should not be the case.

(Rosser, *Logic for Mathematicians* (McGraw-Hill, 1953), p. 207)

And so, in answer to the question that opened this section—"Can we be sure Euclidean geometry is consistent?"—we are forced to say: No.

Notes

[1] *and high-school algebra.* Real Analysis gets its name from its focus on the properties of the so-called "real" numbers—the set of numbers that includes 0, the positive and negative whole numbers (like 12 and −7), the "rational" numbers (ratios of whole numbers, like 3/2 and −5/11), and the "irrational" numbers (like π and −$\sqrt{2}$). "Imaginary" numbers (like $\sqrt{-4} = 2i$, where $i = \sqrt{-1}$) and "complex" numbers (like 3 − 5i) are the only numbers you are likely to have heard of that are *not* real numbers, though even they can be defined *in terms of* real numbers.

[2] *geometric models were constructed.* As we will see in the case of Poincaré's model, these also depend, ultimately, on the consistency of Real Analysis.

[3] *Science and Hypothesis.* Pages 64–68 of the 1905 English translation (Dover reprint, 1952). Poincaré describes a seemingly different but equivalent model on pages 41–43.

[4] *Poincaré's model consists.* The formal presentation of Poincaré's model in this paragraph, while motivated by the preceding story, is independent of the story.

[5] *the arithmetic of the whole numbers.* The other real numbers were defined in terms of the whole numbers. For example, the rational numbers were defined to be "equivalence classes" of pairs of whole numbers in which the second member was nonzero. The crucial discovery was of a method for expressing *ir*rational numbers like $\sqrt{2}$ in terms of whole numbers. See Dedekind's essay "Continuity and Irrational Numbers" (1872; English translation, 1901), reprinted in *Essays on the Theory of Numbers* by Richard Dedekind (Dover, 1963).

[6] *circle.* As it is difficult, using pairs of real numbers, to give a satisfactory interpretation of Euclid's term "line," I have interpreted the term "circle" instead. This is adequate for our purpose because circles are the only "lines," other than straight lines, that Euclid ever uses. In the interpretation the pair (h, k) corresponds to the center of the circle and r to the length of the radii.

[7] *does not use a model.* See for example Chapter V of *Gödel's Proof* by Ernest Nagel and James R. Newman (New York University Press, 1958); or pages 31–37 of *Introduction to Mathematical Logic* by Elliott Mendelson (Van Nostrand, 1979).

[8] *the system itself.* From "Gödel's Proof" by Nagel and Newman, *Scientific American*, June 1956. This article was later enlarged into the book mentioned in the previous note.

[9] *first decade of this century.* The same that we referred to on page 10.

CHAPTER 8

Geometry and the Story Theory of Truth

Kant Revisited

Sooner or later, people who study hyperbolic geometry develop a "hyperbolic" intuition. Pictures stop looking wrong. Theorems can be anticipated. The world can be seen, at will, in either Euclidean or hyperbolic terms.

This is a mortal blow to Kant's doctrine of space. Kant had said that data from our senses is organized along Euclidean lines before we receive it in awareness. If this were the case, it would be impossible to experience the world as hyperbolic. Yet over the last century and a half thousands of people have testified that they have learned to do precisely this, and that furthermore, having done so, have come to regard their older, Euclidean perception of the world as having been learned as well.

"How," Kant had asked (*Prolegomena*, Preamble, §5), "are synthetic propositions *a priori* possible?" His answer, for Euclidean geometry, had been his doctrine of space. With that discredited, Kant's attribution of synthetic a priori status—diamond status—to the Euclidean postulates and theorems was undercut as well.

Today the prevailing view is that Kant, living before the invention of non-Euclidean geometry, had failed to appreciate *how* pure "pure mathematics" really is. Properly, say modern philosophers of mathematics, the traditional designation "pure mathematics"—meaning mathematics *as such*, independent of any application to the world—should only be used in reference to *formal* axiomatic systems (p. 162), because only formal systems are completely divorced from their empirical origins. Since in a formal axiomatic system the primitive terms, and therefore the axioms and theorems, have no meaning, we cannot possibly judge the axioms or theorems to be true (or, for that matter, false). Therefore, while the postulates and theorems of pure geometry are *in a sense* a priori—they are independent of experience—they are neither a priori nor synthetic in *Kant's* sense because for Kant those terms could be applied only to statements we judge to be true.

A variation of this view, due to Poincaré, allows the postulates and theorems of pure geometry to be a priori (and so certain) even in Kant's sense, but concludes they are analytic (and so uninformative). It's true that in pure geometry, because the primitive terms are uninterpreted, we cannot judge the postulates to be true. But they are the only source of information about the primitive terms we have. When we are proving a theorem we have no choice but to treat the postulates, for the sake of the deduction, as if they *were* true. In effect, then, the postulates *define* the primitive terms for us—not explicitly, or completely, but insofar as we need them to be defined to carry out the business at hand. Poincaré said that this is how the postulates of pure geometry *should* be viewed.

The geometrical [postulates] are . . . neither synthetic a priori intuitions nor experimental facts. They are conventions. Our choice among all possible conventions is *guided* by experimental facts; but it remains *free*, and is only limited by the necessity of avoiding every contradiction, and thus it is that postulates may remain rigorously true even when the experimental laws which have determined their adoption are only approximate. In other words, *the [postulates] of geometry . . . are only definitions in disguise.**

From this perspective "points," "straight lines," and pure geometry's other primitive terms *are* defined, implicitly. We understand them, if the geometry is Euclidean geometry, as "things for which the Euclidean postulates are true"; or if the geometry is hyperbolic geometry, as "things for which the hyperbolic postulates are true." With the primitive terms of either geometry so understood we could immediately judge its postulates to be true, but as our judgment would have required nothing beyond that understanding the postulates, and so the theorems, would be analytic.

Just as use of the term "pure mathematics" is now restricted to formal axiomatic systems, use of the traditional companion term, "applied mathematics," is now seen to be appropriate only for material axiomatic systems (i.e., formal axiomatic systems which have been interpreted—see page 6); and there is the additional proviso that *the interpreted postulates must be testable* by the standards of the discipline within whose purview the interpretation is made. This means that any applied geometry would be a part of, say, physics, or astronomy, or some other science. A specific physical interpretation would have to accompany each primitive term. Within physics, for example, one might interpret the term "straight line" as a stretched fiber—as the Egyptians are supposed to have done, and Euclid may have intended in the *Elements*; or as the path of a ray of light—as Euclid explicitly did in the *Optics*, and modern physicists actually do. With every term so interpreted the postulates would become statements about the behavior of material objects, statements that, if

---

* Reprinted from Poincaré, *Science and Hypothesis*, Dover, 1952, p. 50, with permission.

found upon experimentation to be true, even Kant would admit would be empirical. (We "learn from experience," he said on p. 107, "that bodies are heavy, and will fall when their supports are taken away." The interpreted postulates would be qualitatively similar statements.)

In sum: it is possible to classify the postulates of either Euclidean geometry or hyperbolic geometry under Kant's distinctions, but first we have to decide whether we are speaking of the geometry in its pure (formal, uninterpreted) or an applied (material, interpreted) form. And it turns out that, regardless of what we decide, the postulates are never classified as *both* synthetic *and* a priori. The postulates of a pure geometry are either not a priori at all because they are meaningless, or, *à la* Poincaré, they are a priori but only analytic. The postulates of an applied geometry, if found upon experimentation to be true, would be synthetic but only empirical.

While Kant's concept of mathematics is thus rejected by most modern philosophers of mathematics, I hope you will not conclude from this that Kant was some kind of nitwit. He was a scrupulous and profound thinker. The modern position that every mathematical system is *at root* formal, uninter- preted, and so about *nothing in particular*, on which the rejection of Kant is based, was a radical departure from the past, reached in the second half of the nineteenth century, long after Kant was dead, and only after a decades- long reappraisal of mathematics prompted in large measure by the invention of non-Euclidean geometry. Besides, Kant's philosophy contains more than an outmoded concept of mathematics. He was in fact a prophetic figure, whose emphasis on the role of the knower in shaping knowledge (although for Kant it was an unconscious role) paved the way for the emerging epistemology we will soon discuss.

Granting that Kant labored under an enormous handicap, and that, short of inventing non-Euclidean geometry himself, he probably couldn't have made the one distinction (formal axiomatic system *vs.* material axiomatic system) now seen as crucial, we can still ask: *Why*, in retrospect, did the Euclidean postulates strike him as synthetic a priori statements, as diamonds, in the first place?

On page 5 I called the *Elements* a "material" axiomatic system. It is not.

Euclid did, I think, consider his primitive terms to have meanings apart from the relations among them specified in the postulates; they were more than otherwise-unknown "things for which the postulates are true." We have remarked that Euclid's "Definitions" of his primitive terms are strictly not definitions, but rather explanations. In fact, because these "Definitions" employ concepts outside Euclid's system, they resemble what would today be called "interpretations" of the primitive terms. His diagrams heighten our impression that we *know* what "points" and "straight lines" *are*, that they are *things* we can perceive, or almost perceive, in the physical world. It is this *appearance* the *Elements* gives of interpreting its primitive terms, of being a science, that I was acknowledging when I called it a "material" axiomatic system.

The idea of a "material" axiomatic system only came into focus, however, in conjunction with that of a "formal" axiomatic system—in the second half of the nineteenth century, well after science had become experimental.[1] By the standards of experimental science Euclid's "Definitions" of his primitive terms are too *hazy* to be accepted as interpretations. By not referring explicitly to specific physical objects, they do not allow experimental tests of the postulates to be made. Euclid may well have envisioned a "straight line" to be a stretched fiber, or the path of a light-ray, but if so, he did not say so clearly enough for his "Definition" to pass muster today as an interpretation.

The *Elements* is thus revealed as *partially* interpreted, midway between a formal axiomatic system and a material axiomatic system. I think this is why Kant could find within it[2] features of both pure geometry and applied geometry. Because of the extent to which the primitive terms are *not* interpreted—they are not described as corresponding to specific physical objects, no experimental tests of the postulates are proposed or cited—the postulates seem independent of sense-experience and so a priori. But because of the extent to which the primitive terms *are* interpreted—explanations are offered, diagrams provided—the postulates are brought into the company of empirical principles (like "bodies are heavy, and will fall when their supports are taken away"), which are all synthetic.

The Luneburg–Blank Theory of Visual Space

There is no branch of mathematics, however abstract, that will not eventually be applied to the phenomena of the real world.

—Lobachevsky[3]

Underlying Kant's doctrine of space was his belief that the space we experience—in particular, the space we see—is Euclidean. However, research since World War II indicates that binocular visual space—the space we see using both eyes—is not Euclidean, but hyperbolic! If this is correct, the implication is that when we *think* we see the world as Euclidean we are allowing the presumed truth of Euclidean geometry to color the experiences we actually have. There is a "Sense-Data Processor," all right, but it is our past study of Euclidean geometry! While Kant's doctrine of space hardly requires further refutation, the irony of this situation makes a brief discussion irresistable. (The main thread of this chapter continues in the next section.)

The theory that binocular visual space is hyperbolic was proposed in 1947 by Rudolf K. Luneburg and refined after Luneburg's death in 1949 by Albert A. Blank. From a few natural assumptions, later simplified by Blank, Luneburg had decided that the geometry of binocular visual space is either Euclidean, hyperbolic, or elliptic ("elliptic" geometry is another non-Euclidean geometry invented in the 1850s). A series of carefully-conducted experiments overseen by Blank in the 1950s confirmed the hypothesis that the geometry is in fact hyperbolic.

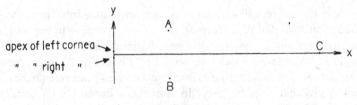

Figure 254

The experimental base for the theory [wrote Blank in 1978] isolates the factor of binocularity. The observer's head is fixed to avoid motion parallax. Because the region of distinct binocular vision is severely restricted in fixed gaze, the more natural scanning mode of observation is used; the observer is asked to actively survey the entire presented configuration in making settings. At best the stimuli are composed of a number of low intensity lights in a totally dark surround. The lights are adjusted to equal sensory brightness and made as small as possible to approximate geometrically ideal points. In this way the well-known monocular depth cues of size, overlay and relative brightness are largely avoided.*

Here is Blank's account of one particularly striking experiment.**

Three starlike lights A, B, C determining the experimental triangle are presented to the observer in the eye-level plane. In that plane a coordinate system is chosen so that the x axis is directed sagittally forward along the median line. The y axis is directed toward the observer's left along a line joining the apexes of his corneas. The coordinates of a point (x, y) will be given in inches and decimal fractions of an inch. For convenience in taking and presenting data the triangle is made symmetric to the median with vertices $C = (108, 0)$, $A = (28, 12)$, $B = (28, -12)$. This does not preclude the use of asymmetric observers. .... For a strongly asymmetric observer the data from right and left sides cannot be pooled. No such observers were found in this series.

I have drawn the configuration to scale, looking down from the ceiling (see Figure 254). The relative positions of lights and observer are typical of these experiments; when the lights are very close to the observer or far off to the side, the situation is more complicated.

The observer is first presented the three lights A, B, C alone; these are kept fixed throughout the experiment. A fourth light is introduced somewhere to the left of the median and the observer is told to ask the experimenter to move the light in order to satisfy the instruction, "Place this light so that you see it as lying on the left side of the triangle exactly equidistant from the two endpoints. In executing this task be sure to look squarely at each light and fix its position carefully rather than superficially glide

---

* A. A. Blank: "Metric Geometry in Human Binocular Perception: Theory and Fact," in Leeuwenberg and Buffart (eds.), *Formal Theories of Visual Perception* (pp. 83–84). Copyright © 1978 by John Wiley & Sons, Ltd. Reprinted with permission of John Wiley & Sons, Ltd.
** A. A. Blank: "Curvature of Binocular Visual Space. An Experiment," in *Journal of the Optical Society of America* (March, 1961), pp. 336–338. Reprinted with permission.

from light to light." The fourth light is then turned off and a fifth light is introduced at the right and the same task is performed on the right side of the triangle. After this initial setting, both lights are turned on and remain on simultaneously; the observer is asked to repeat the bisection of the sides of the triangle a variable number of times. Between the observer's settings the lights are displaced at random so that the observer makes a fresh beginning each time. The experiment is repeated until evidence of any continuing trend is not apparent in the last five or more settings. The medians of the x and y coordinates after the termination of the trend are used as the most convenient representative data. We denote these median data by (x_α, y_α) and (x_β, y_β) for the left and right sides, respectively.

Later in the article Blank explains why early data are discarded. "General experience in the domain of pure binocular observation," he writes, "has demonstrated the need for practice on the part of the observer before he accustoms himself to a situation in which he operates on minimally sufficient clues."

In the table we list the means on the right and left of the representative data, namely,

$$x^* = \tfrac{1}{2}(x_\alpha + x_\beta), \; y^* = \tfrac{1}{2}(y_\alpha - y_\beta).$$

The tolerance indicated is the maximum of the root-mean-square deviations from the median on each side. In almost every case the distance of the medians on the two sides from their mean is definitely less than the maximum root-mean-square tolerance; in those few cases where it is larger, we give this larger figure instead. Also given are the number n of repetitions of the experiment and the number k of settings from which the median is taken.

Observer	Age	γ', γ''		α, β			
		$\bar{y}$	$\bar{x}$	y^*	x^*	n	k
GAH	36	5.58 ± 0.34	28.40 ± 0.31	2.89 ± 0.26	91.00 ± 0.59	20	5
RGB	15	4.32 ± 0.31	28.43 ± 0.40	3.43 ± 0.32	90.69 ± 0.52	15	5
PE	16	4.07 ± 0.45	29.36 ± 0.47	3.93 ± 0.54	73.04 ± 1.19	14	5
RRC	35	3.40 ± 0.40	28.76 ± 0.17	2.14 ± 0.48	94.90 ± 0.98	11	5
MD	15	1.68 ± 0.64	28.15 ± 0.68	5.00 ± 0.57	71.46 ± 1.39	14	14
IG	17	0.80 ± 0.28	28.76 ± 1.00	4.67 ± 0.45	70.50 ± 0.68	11	5
WHF	14	-0.10 ± 0.74	28.34 ± 0.24	3.46 ± 0.39	93.25 ± 0.76	31	7

In the second part of the experiment, two lights are fixed at $\alpha = (x^*, y^*)$ and $\beta = (x^*, -y^*)$. The observer is asked whether these lights satisfy the criterion of the instructions. In no case was the answer negative. [See Figure 255.]

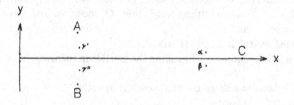

Figure 255

Next he is instructed to set a light on the base AB of the triangle first so that (1) the distance from A to the new light equals that from α to β and then so that (2) the distance from B to the new light is equal to the distance from α to β. Five settings are made alternately under each of these instructions. The medians of the coordinates of the set points are used to determine points

$$\gamma' = (x', y'), \qquad \gamma'' = (x'', y'')$$

which serve as representative data. The sensory relation assumed to hold for the interpretation of the experiment is

$$\alpha\beta = A\gamma' = \gamma''B.$$

The means

$$\bar{x} = \tfrac{1}{2}(x' + x'') \quad \text{and} \quad \bar{y} = \tfrac{1}{2}(y' - y'')$$

and their tolerances (computed as for α and β) are given in [the table].

Observer GAH, for example, set the midpoints of AC and BC at about $\alpha = (91, 2.89)$ and $\beta = (91, -2.89)$, positions computed from statistical averages of the last 5 of his 20 bisections of the sides. (Euclidean midpoints would be $(68, 6)$ and $(68, -6)$.) In the second part of the experiment he set points on AB whose distances from A and B are equal to $\alpha\beta$ at about $(\bar{x}, \bar{y}) = (28.4, 5.58)$ and $(\bar{x}, -\bar{y}) = (28.4, -5.58)$, positions based—as for every observer—on 5 pairs of settings. (Euclidean settings of the points would coincide at $(28, 0)$.)

The evidence that GAH's binocular visual space is hyperbolic is based on the fact that the length of the straight line joining the midpoints of two sides of a triangle is equal to, less than, or greater than half the third side according as the geometry is Euclidean, hyperbolic, or elliptic. (The hyperbolic case is Chapter 6, Exercise 5.)

The observers are listed in decreasing order of $\bar{y}$, the number that indicates, on the whole, how much shorter than $\tfrac{1}{2}AB$ they perceived $\alpha\beta$ to be. The conclusion is that the binocular visual space of six of the seven observers is significantly hyperbolic. Blank comments on the seventh observer.

The performance of the Euclidean observer WHF is distinctive enough to report in detail. As with a number of other observers, his settings exhibited a marked initial trend. ... The only specially remarkable feature about this trend is its length. The first 18 settings were taken in one session and the conclusion of the experiment was put off to a later day when 13 more settings were taken. Of these, the last seven were used to obtain the medians α and β.

Upon hearing instruction 2 WHF asked,

"Isn't there a theorem here?" (meaning the Euclidean theorem, of course) and was told

"Forget the theorem and go by what you feel directly."

"I can't forget the theorem."

The Diamond Theory in Decline

In Chapter 3, I called the belief that

(1) Diamonds—informative, certain truths about the world—exist

the "Diamond Theory" of truth. I said that for 2200 years the strongest evidence for the Diamond Theory was the widespread perception that

(2) The theorems of Euclidean geometry are diamonds.

Hyperbolic geometry does not refute the Diamond Theory. To my mind the Diamond Theory *cannot* be refuted. Refuting the Diamond Theory would mean *proving* that diamonds don't exist, which would make the statement "Diamonds don't exist" a diamond!

Hyperbolic geometry does refute (2), however. By providing an alternative way of looking at geometric figures that is both internally consistent and compatible with everyday experience, it robs the Euclidean theorems of their certainty. Hyperbolic geometry thus deprives the Diamond Theory of its principal psychological support, making the Diamond Theory much harder to believe.

As the news about non-Euclidean geometry spread—first among mathematicians, then among scientists and philosophers—the Diamond Theory began a long decline that continues today.

Factors outside mathematics have contributed to this decline. Euclidean geometry had never been the Diamond Theory's *only* ally. In the eighteenth century other fields had seemed to possess diamonds, too; when many of these turned out to be man-made, the Diamond Theory was undercut. And unlike earlier periods in history, when intellectual shocks came only occasionally, received truths have, since the eighteenth century, been found wanting at a dizzying rate, creating an impression that perhaps *no* knowledge is stable.

Other factors notwithstanding, non-Euclidean geometry remains, I think, for those who have heard of it, the single most powerful argument against the Diamond Theory—first, because it overthrows what had always been the strongest argument in favor of the Diamond Theory, the objective truth of Euclidean geometry; and second, because it does so *not* by showing Euclidean geometry to be false, but by showing it to be merely uncertain. If the outcome of the non-Euclidean revolution had been that Euclidean geometry is false—and hyperbolic geometry (or some other non-Euclidean geometry) is true—we would be tempted to respond as people faced with scientific revolutions so often have: "Yes, we were wrong before; but now we know the truth." The non-Euclidean revolution does not allow this response, as it leaves the question of which geometry is true open. I can think of every triangle as having angle-sum 180°, if I like, without fear of logical inconsistency or conflict with everyday experience; but I can also think of every triangle as having angle-sum *less* than 180°, if I prefer, and still have no fear of inconsistency or conflict with everyday experience. Hyperbolic geometry is more complicated than Euclid's, but *it is an equally valid description of everyday experience*. This suggests that "the truth" in other domains may be multiple, too, in which case diamonds don't exist.

The Story Theory of Truth

... the classical mind says, that's only a story, but the modern mind says, there's only story.

—John Dominic Crossen, *The Dark Interval**

A new epistemology is emerging to replace the Diamond Theory of truth. I will call it the "Story Theory" of truth:

There are no diamonds. People make up stories about what they experience. Stories that catch on are called "true."

The Story Theory of truth is itself a story that is catching on. It is being told and retold, with increasing frequency, by thinkers of many stripes. Here are some expressions of it I ran across while writing this book.

This is from an essay by Robert Frost.

Poetry is simply made of metaphor. So also is philosophy—and science, too, for that matter, if it will take the soft impeachment from a friend.

—"The Constant Symbol," in Cox and Lathem (eds.),
Selected Prose of Robert Frost (Collier, 1968), p. 24

This is from a book about modern physics.

... when we try to understand nature, we should look at the phenomena as if they were *messages* to be understood. Except that each message appears to be random until we establish a code to read it. This code takes the form of an abstraction, that is, we choose to ignore certain things as irrelevant and we thus partially select the content of the message by a free choice. These irrelevant signals form the "background noise," which will limit the accuracy of our message.

But since the code is not absolute there may be several messages in the same raw material of the data, so changing the code will result in a message of equally deep significance in something that was merely noise before, and *conversely*: In a new code a former message may be devoid of meaning.

Thus a code presupposes a free choice among different, complementary aspects, each of which has equal claim to *reality*, if I may use this dubious word.

Some of these aspects may be completely unknown to us now but they may reveal themselves to an observer with a different system of abstractions.

---

* From *The Dark Interval* by John Dominic Crossen. © 1975 Argus Communications, a division of DLM, Inc., Allen, TX 75002. Reprinted with permission.

But ... how can we then still claim that we *discover* something out there in the objective real world?

—J. M. Jauch, *Are Quanta Real?**

These are from a philosophical book about theology.

I propose, then, to consider as most interesting the story that art and science, or poetic intuition and scientific achievement, are not two simultaneous and separate ways of knowing but two successive and connected moments of all human knowledge; ... and that "reality" is the world we create in and by our language and our story so that what is "out there," apart from our imagination and without our language, is as unknowable as, say, our fingerprints, had we never been conceived. To ask, in other words, what is "out there" apart from the story in which "it" is envisioned, should strike us as strangely as would the question of how one might feel today about the fact that one had never been born. I am not saying we cannot know reality. I am claiming that what we know *is* reality, *is* our reality here together and with each other.

Reality is neither *in here* in the mind nor *out there* in the world; it is the interplay of both mind and world in language.

—John Dominic Crossen, *The Dark Interval***

This is from a book about the history of the English language.

When some people argue that histories should tell the truth, they usually mean by truth the straightforward, unadorned facts of what happened. They believe that historians who select, arrange, and shape their data to make a point about the past or who use the past to prove something about the present are not being objective, and, hence, not telling the "truth." History, they believe, is properly the objective facts the historians can recover from the past arranged just as they occurred to relate what really happened.

Such histories have never existed and never will. The individual mind of the historian, shaped by his times, influenced by his theory of history, and controlled by his unique personal character, must always stand between the leavings of the past and the work that represents his understanding of how those leavings reveal what happened and why. Just choosing what to write about reveals the mind of the historian evaluating and selecting from all the events of the past only those most important to him.

. . .

At its simplest and most elemental level, historians create a past because their community wants to know where it came from and how things got to be the way they are. The earliests myths, the stories of the most primitive peoples about their gods, are a kind of history that explains how the universe and the earth came about—why the sun and the moon, why the animals, why man. The voyage of Mariner 10 beyond Jupiter and out of our solar system represents our curiosity about the origins of the universe. The Bible was once unquestioned history for most Christians. For many it still is. So once for some were the legends of the Plains Indians, and the Greek myths and *Beowulf*

---

* Reprinted from J. M. Jauch, *Are Quanta Real?* (Indiana University Press, 1973) (p. 64), with permission.
** From *The Dark Interval* by John Dominic Crossen. © 1975 Argus Communications, a division of DLM, Inc., Allen, TX 75002. Reprinted with permission.

and the story of Valley Forge, Gettysburg, Lewis and Clark, and the Alamo. We no longer call all of these history because our age has different criteria for accepting some stories rather than others as satisfying ways to organize and explain our past.

· · ·

We must understand that the data [the historian] selects and how he arranges them depend on his reasons for writing history in the first place, that "truth" depends finally on how his theories interpret what the interests and values of an age allow him to recognize as the truth.

—Joseph M. Williams,
*Origins of the English Language**

And this is from a best-selling novel.

I see us all sitting around naked, shivering, in a huge circle, looking up as the sky turns black and the stars flare out and somebody starts to tell a story, claims to see a pattern in the stars. And then someone else tells a story about the eye of the hurricane, the eye of the tiger. And the stories, the images, become the truth and we will kill each other rather than change one word of the story. But every once in a while, someone sees a new star, or claims to see it, a star in the north that changes the pattern, and that is devastating. People are outraged, they start up grunting in fury, they turn on the one who noticed it and club her to death. They sit back down, muttering. They take up smoking. They turn away from the north, not wanting it to be thought that they might be trying to catch sight of her hallucination. Some of them, however, are true believers, they can look straight north and never see even a glimpse of what she pointed to. The foresighted gather together and whisper. They already know that if that star is accepted, all the stories will have to be changed. They turn suspiciously to sniff out any of the others who might surreptitiously turn their heads to peer at the spot where the star is supposed to be. They catch a few they think are doing this; despite their protests, they are killed. The thing must be stopped at the root. But the elders have to keep watching, and their watching convinces the others that there's really something there, so more and more people start to turn, and in time everyone sees it, or imagines they see it, and those who don't claim they do.

So earth feels the wound, and Nature from her seat, sighing through all her Works, gives signs of woe, that all is lost. The stories all have to be changed: the whole world shudders. People sigh and weep and say how peaceful it was before in the happy golden age when everyone believed the old stories. But actually nothing whatever has changed except the stories.

I guess the stories are all we have, all that makes us different from lion, ox, or those snails on the rock.

—Marilyn French, *The Women's Room***

The Diamond Theory and Story Theory are alternative ways of looking at what we humans are doing when we come to "know" something.

My own viewpoint is the Story Theory. I cannot "prove" it is "right," or even analyze my own reasons for accepting it; I can only report that the Story Theory describes what I have seen.

In college I was prepared to adopt the Diamond Theory. I was excited at the possibility that people had discovered extracts, so to speak, from God's blueprint for the universe. And I thought: if humanity is in possession of any objective truth at all, surely there must be some measure of objective truth in mathematics, or perhaps science or philosophy.

I've since spent twenty years studying mathematics, science, and philosophy, and I concluded long ago that each enterprise contains only stories (which the scientists call "models of reality"). I had started by hunting diamonds; I did find dazzlingly beautiful jewels, but always of human manufacture.

I am not disappointed. I would rather view people as active creators of truth than as passive searchers (for all the ingenuity the search may involve) after truth we have no part in shaping. The gain in human freedom and dignity is exhilirating. We have a role in the creation of the world! We are like the mythmakers before Thales, except that, naturally, we feel *our* stories are better.

Notes

[1] *after science had become experimental.* While the Greeks certainly made observations, sometimes in great detail, they had no concept of "experiments" in the modern sense.

[2] *Kant could find within it.* While Kant seems not to have used the *Elements* itself, in the book he did use primitive terms were undoubtedly "defined" and represented in diagrams.

[3] *Lobachevsky.* Quoted in the *American Mathematical Monthly* (February 1984), p. 151.

Bibliography

This is not a complete list of books consulted, or even cited. It *is*, I hope, a starting-place for anyone who has found parts of this book interesting and would like to read more.

Historical, Philosophical

Stephen F. Barker, *Philosophy of Mathematics* (Prentice-Hall, 1964).

Roberto Bonola, *Non-Euclidean Geometry: A Critical and Historical Study of Its Development* (1906), translated by H. S. Carslaw (1911; Dover reprint, 1955). The Dover edition includes János Bolyai's *Absolute Science of Space* (1832) and Nicolai Lobachevsky's *Geometrical Researches on the Theory of Parallels* (1840), both translated by George B. Halsted.

Ettore Carruccio, *Mathematics and Logic in History and in Contemporary Thought*, translated by Isabel Quigley (London: Faber and Faber, 1964).

John Dominic Crossen, *The Dark Interval: Towards a Theology of Story* (Argus Communications, 1975).

Philip J. Davis and Reuben Hersh, *The Mathematical Experience* (Birkhäuser Boston, 1981).

Kenneth Dover, *The Greeks* (University of Texas Press, 1981).

Henri Frankfort and others, *Before Philosophy: the Intellectual Adventure of Ancient Man* (Penguin, 1949).

Hans Freudenthal, "The Main Trends in the Foundations of Geometry in the 19th Century," in Nagel, Suppes, and Tarski (editors), *Logic, Methodology and Philosophy of Science* (Stanford University Press, 1962), pp. 613–621.

Rom Harré, *The Anticipation of Nature* (London: Hutchinson, 1965).

Sir Thomas L. Heath, *A History of Greek Mathematics* (1921; Dover reprint, 2 volumes, 1981).

Morris Kline, *Mathematics: the Loss of Certainty* (Oxford University Press, 1980).

Wilbur R. Knorr, *The Evolution of the Euclidean Elements* (D. Reidel, 1975).

Robert M. Pirsig, *Zen and the Art of Motorcycle Maintenance: An Inquiry into Values* (Bantam, 1975).

Henri Poincaré, *Science and Hypothesis* (1902), translated by W. J. G. (1905; Dover reprint, 1952).

261

W. V. Quine and J. S. Ullian, *The Web of Belief* (Random House, 1970).

J. Barkley Rosser, *Logic for Mathematicians* (McGraw-Hill, 1953; Chelsea reprint, 1978).

Bruno Snell, *The Discovery of the Mind in Greek Philosophy and Literature*, translated by T. G. Rosenmeyer (1953; Dover reprint, 1982).

Paul Van Buren, *The Edges of Language* (Macmillan, 1972).

Technical

János Bolyai, *Absolute Science of Space* (1832), translated by George B. Halsted. See Bonola.

Howard Eves, *A Survey of Geometry* (Allyn and Bacon, 1972).

Richard L. Faber, *Foundations of Euclidean and Non-Euclidean Geometry* (Marcel Dekker, 1983).

Marvin Jay Greenberg, *Euclidean and Non-Euclidean Geometries: Development and History* (W. H. Freeman, 1980).

Hardy, Rand, Rittler and Blank, Boeder, *The Geometry of Binocular Space Perception* (Knapp Memorial Laboratories, Institute of Ophthalmology, Columbia University College of Physicians and Surgeons, 1953). The report of the laboratory where the primary investigations of the Luneburg–Blank theory of visual space were conducted. For later developments and a short bibliography, see pp. 83–102 of the book edited by Leeuwenberg and Buffart I cited on p. 252.

[Heath's *Euclid*] Sir Thomas L. Heath, *The Thirteen Books of Euclid's Elements* (1908, 1925; Dover reprint, 3 volumes, 1956). The *Elements* translated from the definitive Greek text, with an extensive introduction and commentary. The translation alone, without the introduction or commentary, is also in volume 11 of "The Great Books of the Western World" (Encyclopaedia Britannica, 1952), available in many public libraries.

David Hilbert, *Foundations of Geometry*, 10th edition (1968), translated by Leo Unger (Open Court, 1971).

Nicolai I. Lobachevsky, *Geometrical Researches on the Theory of Parallels* (1840), translated by George B. Halsted. See Bonola.

George E. Martin, *The Foundations of Geometry and the Non-Euclidean Plane* (Springer-Verlag, 1975).

Edwin E. Moise, *Elementary Geometry from an Advanced Standpoint* (Addison-Wesley, 1974).

Gerolamo Saccheri, *Euclides Vindicatus* (1733), translated by George B. Halsted (Chelsea reprint, 1986).

Harold E. Wolfe, *Introduction to Non-Euclidean Geometry* (Holt, Rinehart and Winston, 1945).

Index